Removal of Organic Pollution in Water Environment

Removal of Organic Pollution in Water Environment

Special Issue Editors

Joanna Karpińska
Urszula Kotowska

MDPI • Basel • Beijing • Wuhan • Barcelona • Belgrade

MDPI

Special Issue Editors
Joanna Karpińska
Faculty of Chemistry,
University of Bialystok
Poland

Urszula Kotowska
Faculty of Chemistry,
University of Bialystok
Poland

Editorial Office
MDPI
St. Alban-Anlage 66
4052 Basel, Switzerland

This is a reprint of articles from the Special Issue published online in the open access journal *Water* (ISSN 2073-4441) in 2019 (available at: https://www.mdpi.com/journal/water/special_issues/organic_pollution).

For citation purposes, cite each article independently as indicated on the article page online and as indicated below:

LastName, A.A.; LastName, B.B.; LastName, C.C. Article Title. *Journal Name* **Year**, *Article Number, Page Range.*

ISBN 978-3-03921-840-0 (Pbk)
ISBN 978-3-03921-841-7 (PDF)

Contents

About the Special Issue Editors

Joanna Karpinska is currently Full Professor and Head of the Environmental Chemistry Research Group in the Department of Analytical and Inorganic Chemistry in the Faculty of Chemistry at the University of Bialystok. Her current research interest is the study of processes determining the durability of selected biologically active compounds in the environment and new photocatalysts.

Urszula Kotowska works as an Assistant Professor in the Department of Analytical and Inorganic Chemistry in the Faculty of Chemistry at the University of Bialystok. Her scientific interests concern the determination of organic micropollutants in water samples and the search for new solutions to remove these impurities. She is a specialist in gas chromatography, mass spectrometry and micro extraction techniques. She has served as a reviewer of scientific journals in the field of analytical chemistry and environmental research, and has also worked as a guest editor and a member of the organizing committee of international scientific conferences.

water

MDPI

Editorial

Removal of Organic Pollution in the Water Environment

Joanna Karpińska and Urszula Kotowska *

Institute of Chemistry, University of Bialystok, Ciolkowskiego 1K Street, 15-245 Bialystok, Poland;
joasia@uwb.edu.pl
* Correspondence: ukrajew@uwb.edu.pl; Tel.: +48-85-738-81-11; Fax: +48-85-747-01-13

Received: 8 September 2019; Accepted: 26 September 2019; Published: 28 September 2019

check for
updates

Abstract: The development of civilization entails a growing demand for consumer goods. A side effect of the production and use of these materials is the production of solid waste and wastewater. Municipal and industrial wastewater usually contain a large amount of various organic compounds and are the main source of pollution of the aquatic environment with these substances. Therefore, the search for effective methods of wastewater and other polluted water treatment is an important element of caring for the natural environment. This Special Issue contains nine peer-review articles presenting research on the determination and removal of environmentally hazardous organic compounds from aqueous samples. The presented articles were categorized into three major fields: new approaches to the degradation of water pollutants, new methods of isolation and determination of the emerging organic contaminants (EOCs), and the occurrence of EOCs in the water environment. These articles present only selected issues from a very wide area, which is the removal of organic pollution in water environment, but can serve as important references for future studies.

Keywords: emerging organic contaminants; water environment; EOCs determination; wastewater purification; advanced oxidation processes; electrochemical degradation; biosorption; liquid-liquid continuous extraction; fractional distillation

1. Introduction

Water is essential for life, and although approximately 70% of the Earth' surface is covered with water, only a small fraction (2.5%) is freshwater compatible with terrestrial life. Nowadays, a continuous increase in water demand is observed as a consequence of demographic growth, industry demand, and living conditions. At present, the societies of developed countries are aware of the importance of water quality, especially in western countries. It is a matter of concern that half of the European countries are already facing water stress. According to the European Environmental Agency Report, only around 40% of surface waters (rivers, lakes, and transitional and coastal waters) are in good ecological status or potential, and 38% are in good chemical status [1]. An intensive use of chemicals in everyday activities and unrestricted access to medicines has resulted in increased waste production and an intense emission of typical as well as new organic compounds into the surrounding environment. In recent years, the newly occurred compounds, called emerging organic contaminants (EOCs), are becoming more and more observable. This is very heterogeneous group of substances containing compounds from various chemical groups. They are created by natural as well as anthropogenic compounds with a presence that was not previously detected due to the lack of sensitive analytical methods, or their adverse health effects were not known. Their impact on living organisms is in general unknown, but the provided experiments have proved their negative influence on vitality, life span, and reproductive success [2,3]. Some of them exhibit disrupting endocrine effects or are suspected to cause it. This group of compounds is distinguished by a separate group named the endocrine disrupting compounds (EDCs) [4]. Their presence in the environment arouses special concern because they change the hormonal equilibrium not only of wild organisms but also of humans. The EDCs are very ubiquitous in every element of environment such as surface and ground waters, soil, and air. Despite their presence at low concentrations, they are considered as persistent due to their continuous delivery to the environment.

A number of EOC emission sources have been identified, but discharges of effluents from wastewater treatment plants (WWTPs) are considered as the main ones. These substances are present in all tested wastewater, both before and after the treatment process, usually in concentrations ranging from ng/L to µg/L [5]. The composition and concentration of EOCs in waters supplied to WWTP depend mainly on the socioeconomic characteristics of the population from which wastewater is collected. The concentration of EOCs in the effluents after the purification process depends on both the level of pollution of the incoming waters and the course of the purification process. The concentration ranges of selected groups of EOCs in raw and treated wastewater are presented in Table 1.

Table 1. Concentrations of compounds from the main groups of emerging organic contaminants (EOCs) recorded in urban wastewater and the efficiency of their removal in conventional wastewater treatment plants (WWTP) (based on ref [5–11]).

EOC Group	Concentration Range		Removal Range (%)
	Influent (ng/L)	Effluent (ng/L)	
Hormones	<MQL*– 670	<MQL–275	0–100
Plasticizers	<MQL–5850	<MQL–1840	32–100
Insect repellents	<MQL–42334	<MQL–1663	27–100
UV filters	<MQL–7800	<MQL–772	0–97.5
Surfactants	<MQL–8520	<MQL–3200	42–99
Antimicrobials	<MQL–8880	<MQL–5860	0–100
NSAIDs	<MQL–611000	<MQL–62000	0–100
Antibiotics	<MQL–303500	<MQL–37000	0–100

* MQL—Method Quantification Limit.

The fate and effectiveness of removing organic pollutants are closely related to their physicochemical properties (Henry's constant (H), n-octanol/water partition coefficient (K_{ow}), sorption coefficient (K_D), partition coefficient between soil organic carbon and water (K_{OC})) [12]. Traditional municipal WWTP applied two stages of treatment: mechanical and biological; the third stage with advanced technologies is rarely used. The purpose of the mechanical stage is to remove suspended matter by filtration, sedimentation, and flotation processes. In this stage, the adsorption of pollutants on the suspension particles and absorption in the fats present in the wastewater takes place. During mechanical treatment, hydrophobic compounds (log K_{ow} > 3) undergo partial removal from wastewater, while the efficiency of hydrophilic compound expurgation is very low [13,14]. Biological treatment is usually carried out using the conventional activated sludge (CAS) method under aerobic and anoxic conditions. Removal of organic pollutants at this stage is associated with biodegradation (biotransformation) and sorption on activated sludge [12]. Biodegradation can occur through metabolism or co-metabolism. In order for metabolism to be possible, compounds must have low toxicity, and their concentration must be high enough to support the life processes of microorganisms [15,16]. Co-metabolism is the degradation of organic pollutants by microorganisms when using other substances as a source of nutrients [17]. It is the basic mechanism of biodegradation of EOCs due to their very low concentrations in wastewater. Acid and lipophilic compounds are biodegradable to a much greater extent than polar, neutral, and basic compounds [18]. Removal of organic pollutants from municipal wastewater can also take place by means of volatilization, i.e., transformation from the form dissolved in water to gaseous. The intensity of this process depends on the H value and WWTP operating conditions (intensity of aeration and mixing of wastewater, temperature, and atmospheric pressure).

Conventional wastewater treatment methods are usually not sufficiently effective in removing EOCs. There is demand for the use additional processes like precipitation and chemical coagulation, flocculation, desorption, neutralization, and reverse osmosis for more thorough purification. Unfortunately, these methods, in many cases, are not sufficiently effective in eliminating pollution, or the economical side of carrying out prevents them from being used on a larger scale wastewater treatment. Additionally, the application of physico-chemical processes causes a transfer of EOCs from the water phase to the receiving material or solid phase, which are new wastes, the management of which creates new environmental problems. Another approach is applied by the so called advanced oxidation processes (AOPs) [19]. They are based on the oxidation of organic pollutants by strong oxidants, mainly hydroxyl radicals generated by ozone, hydrogen peroxide, and others [19]. The best oxidation results are achieved in synergistic processes using systems consisting of two or three components, e.g., H_2O_2/UV or O_3/H_2O_2/UV [19,20]. The use of such systems improves the cleaning effects by increasing the efficiency of mineralization of EOCs and reducing the amount of products of incomplete oxidation [21].

2. Overview of the Special Issue

The Special Issue consists of nine papers describing a wide spectrum of research related to the removal of environmentally significant pollutants from aqueous samples and their determination in various matrices. The presented articles can be classified into three broad thematic sectors related to the topic of the special issue: new methods of isolation and determination of compounds from the EOCs group [22–24], occurrence of EOCs in the water environment [24,25], and new approaches to the degradation of significant water pollutants or their removal [26–30].

2.1. New Methods of EOCs Isolation and Determination

Mielech-Łukasiewicz and Starczewska [22] proposed a new electrochemical method for determining the pharmaceutical residues in aqueous samples. The target pharmaceuticals were two compounds from the fungicide group: itraconazole and posaconazole. Cyclic voltammetry and square wave voltammetry with the use of boron-doped diamond (BDD) electrode were used for determining the properties of analytes and in their analytical characterization. The developed method is simple, fast, and sensitive, and its significant advantage is that there is no need to isolate the analytes from the matrix before the determination. The research carried out for river water and tap water samples indicate that the proposed method can be used in the analysis of environmental samples as an alternative to chromatographic methods, which are most often used in EOC determination in natural waters and wastewater [5]. Hryniewicka et al. [23] described the use of high-performance liquid chromatography with ultraviolet detection (HPLC-UV) for the determination of two pharmaceuticals, budesonide and sulfasalazine, in water and wastewater. For the isolation of target compounds from aqueous samples, dispersive liquid-liquid microextraction (DLLME) followed by solidification of floating organic droplet (SFOD) was used. The paper presents the optimization of extraction parameters, such as the type of extraction solvent, pH, and sample ionic strength as well as extraction time. Analysis carried out for spiked samples of river water and municipal wastewater confirmed the usefulness of the proposed method in aquatic environment research. A new, simple, and sensitive method for determination of three hormones (β-estradiol, estrone, and diethylstilbestrol) and ten other compounds from the EOC group (diclofenac, triclosan, propylparaben, butylparaben, benzophenone, 3-(4-methylbenzylidene)camphor, N,N-diethyltoluamide (DEET), bisphenol A, 4-t-octylphenol, and 4-n-nonylphenol) was proposed by Kotowska et al. [24]. Isolation of analytes by ultrasound-assisted emulsification microextraction with solidification of floating organic droplet (USAEME-SFOD) was done simultaneously with derivatization with acetic anhydride to enable determination of EOCs using gas chromatography mass spectrometry (GC-MS). Good accuracy and precision as well as high sensitivity of the developed method enabled its use for natural water samples.

2.2. Occurrence of EOCs in the Environment

In addition to the new method, Kotowska et al. [24] present the results of the determination of thirteen EOCs in groundwater collected at municipal solid waste (MSW) landfill sites and in groundwater from wells distant from sources of pollution. Ten compounds were detected in groundwater from MSW monitoring wells. Five compounds were detected in shallow groundwater wells (depth: 3–8 m), and two compounds in deep drilling wells (depth: 15–46 m). Ferrari et al. [25] described in their paper a well-documented study of the occurrence and concentration in aquatic sediments of ten congeners of polybrominated diphenyl ethers (PBDEs). PBDEs belong to the group of flame retardants and are used for reducing the risk of fires. These compounds are persistent organic pollutants from the EOC group and are toxic to living organisms, including humans. In the presented study, the isolation of PBDEs from sediment samples was done by ultrasound-assisted solvent extraction followed by gas chromatography with electron capture detection (GC-ECD). Six out of ten target compounds were detected in sediment samples taken from Guarani Aquifer in Brazil in concentrations ranging from 0.24 to 2.7 ng/g. According to the authors, pollution of the examined water reservoir with compounds from the PBDEs group is associated with improper management of solid municipal waste.

2.3. New Approaches to Degradation of Water Pollutants

The removal of platinum and palladium from environmental samples by biosorption on fungi *Aspergillus* sp. and yeast *Saccharomyces* sp. was described by Godlewska-Żyłkiewicz et al. [26]. The introduction of these metals into the aquatic environment is mainly associated with the production, use, and recycling of automotive catalysts. Optimization of biosorption parameters such as solution pH, biosorbent mass, and contact time of the solution with the extraction medium was performed to determine the conditions in which the sorption efficiency is highest. The sorption kinetics was tested, and the Langmiur and Freundlich adsorption isotherms were used for interpretation of the process equilibrium. The research conducted shows that the proposed microorganisms can be successfully used to remove platinum and palladium from contaminated waters and industrial waste. Ambauen et al. [27] investigated electrochemical oxidation of the organic model pollutant salicylic acid. Two anode materials, platinum and boron doped diamond, were used along with chloride and sulfate electrolytes. The work presents a detailed kinetics analysis and identification of oxidation process products. Studies have shown that the products of salicylic acid electrochemical oxidation are hydroxylated and chlorinated intermediates, and the dominance of one of the forms depends on the composition of the electrolyte used. The best results of electrochemical degradation of salicylic acid were achieved where the combination of BDD electrode and chloride electrolyte was used, and the worst results were achieved when a platinum electrode was placed in the same electrolyte. A very low oxidation efficiency of the test compound was observed when the sulfate electrolyte was used in combination with both BDD and platinum electrodes. Karpinska et al. [28] describes detailed studies on the kinetics of the degradation processes of doxazosin (DOX) under the influence of sunlight in environmental conditions and some advanced oxidation processes (AOPs). Doxazosin is a biologically active compound used for the treatment of some prostate complaints and hypertension. The authors checked DOX photochemical behaviors and stated that it is a photoliable compound and its degradation is a result of a direct photolysis. Its $t_{1/2}$ in the presence of a natural matrix lasted from 1 h 30 min to 40 min depending on the chemical composition of the samples of surface water. The studies on DOX behavior under the influence of examined chemical and photo-chemical processes (UV/H_2O_2, Fenton and photo-Fenton process, and $SO_4^{\cdot-}$ radicals) were performed. It was stated that $SO_4^{\cdot-}$ radicals are most efficient and caused DOX degradation in a very short time. The application of a new photocatalyst for the degradation of a selected organic compound was proposed by Regulska et al. [29]. The authors examined the photochemical properties of crystalline $NiAl_2O_4$ decorated with graphene quantum dots. They characterized morphology and structure of a synthetized composite using thermogravimetric methods as well as spectral techniques (XRD, ATR-FTIR, SEM, EXD, and UV-Vis diffuse reflectance spectra). Its photocatalytic activity in ratio to chosen model pollutants (rhodamine B, quinolone yellow, eriochrome black, methylene blue, phenol, and thiran) was studied. It was stated that newly obtained material exhibits photocatalytic activity under the influence of visible light. The detailed mechanism of its operation was proposed and discussed. Its efficiency strongly depends on the presence of electron and hole scavengers and the chemical properties of adsorbed organic compounds. The above articles [27–29] concern the application of chemical processes for removal of organic pollutants. Another approach to wastewater cleaning was proposed by Mendoza et al. [30]. The authors focused on the problem of cleaning wastewater generated by the petroleum industry. They proposed the use of continuous liquid-liquid extraction with dichloromethane (CLLE$_{DCM}$) and high-power fractional distillation (HPFD) to resolve this problem. The efficiency of CLLE$_{DCM}$ and HPFD was examined individually and in combination: CLLE$_{DCM}$-HPFD and HPFD-CLLE$_{DCM}$. The chemical parameters of wastewaters were checked. It was stated that all processes remarkably improved the quality of the samples used. The greatest achievements were obtained by HPFD.

3. Conclusions

The presented Special Issue concerns the problem of the appearance of new organic pollutants in surface water bodies. At present, it is obvious that the chemical composition of the surface water is the result of industrial, agriculture, as well as domestic activities of human population. As it mentioned in the introduction, the wastewater treatment plants have been identified as the one of the main sources of organic pollutions. Thus, much more effort should be made for the improvement of wastewater cleaning processes. The authors involved in the preparation of this Special Issue described the results of examinations, at a laboratory scale, of the efficiency of chemical as well as physical processes for the removal or degradation of selected model pollutants. However, it should be noted that extension of the proposed processes for technological scale requires intense additional studies. The environmental studies, especially those concerning the determination of trace impurities, require effective isolation and concentration procedures. The reagents used for this purpose should meet the requirements of green chemistry. The DLLME-SFOD as well as USAEME-SFOD procedures described in this Special Issue seem to be proper for environmental studies as they are effective and environmentally friendly. Another approach is based on the use of BDD electrodes for direct determination of the target analyte in environmental samples. The described method allowed an assay of examined pharmaceuticals without their isolation from liquid samples.

Author Contributions: The authors made fairly equal contributions to this paper. J.K. drafted Section 3 and parts of Sections 1 and 2. U.K. drafted the Abstract and other parts of Sections 1 and 2. Both authors made revisions and edits of the manuscript.

Funding: This research received no external funding.

Acknowledgments: The authors, who served as the guest-editors of this Special Issue, highly appreciate the journal editors, all authors submitting papers, and the anonymous reviewers for their valuable comments.

Conflicts of Interest: The authors declare no conflict of interest. The funders had no role in the design of the study; in the collection, analyses, or interpretation of data; in the writing of the manuscript, or in the decision to publish the results.

References

1. European Environment Agency. *European Waters Assessment of Status and Pressures 2018*; EEA: Copenhagen, Denmark, 2018.
2. Bókony, V.; Üveges, B.; Verebélyi, V.; Ujhegyi, N.; Móricz, Á.M. Toads phenotypically adjust their chemical defences to anthropogenic habitat change. *Sci. Rep.* **2019**, *9*, 3163–3174. [CrossRef] [PubMed]
3. Peteffi, G.P.; Fleck, J.D.; Kael, I.M.; Rosa, D.C.; Antunes, M.V.; Linden, R. Ecotoxicological risk assessment due to the presence of bisphenol A and caffeine in surface waters in the Sinos River Basin—Rio Grande do Sul—Brazil. *Braz. J. Biol.* **2019**, *79*, 712–721. [CrossRef] [PubMed]

4. Kumar, A.; Xagoraraki, I. Pharmaceuticals, personal care products and endocrine-disrupting chemicals in U.S. surface and finished drinking waters: A proposed ranking system. *Sci. Total Environ.* **2010**, *408*, 5972–5989. [CrossRef] [PubMed]

5. García-Córcoles, M.T.; Rodríguez-Gómez, R.; de Alarcón-Gómez, B.; Çipa, M.; Martín-Pozo, L.; Kauffmann, J.-M.; Zafra-Gómez, A. Chromatographic methods for the determination of emerging contaminants in natural water and wastewater samples: A review. *Crit. Rev. Anal. Chem.* **2019**, *49*, 160–186. [CrossRef] [PubMed]

6. Tran, H.; Reinhard, M.; Gin, K.Y.-H. Occurrence and fate of emerging contaminants in municipal wastewater treatment plants from different geographical regions—A review. *Water Res.* **2018**, *133*, 182–207. [CrossRef] [PubMed]

7. Cantero, M.; Rubio, S.; Pérez-Bendito, D. Determination of alkylphenols and alkylphenol carboxylates in wastewater and river samples by hemimicelle-based extraction and liquid chromatography–ion trap mass spectrometry. *J. Chromatogr. A* **2006**, *1120*, 260–267. [CrossRef] [PubMed]

8. Hernando, M.D.; Mezcua, M.; Gómez, M.J.; Malato, O.; Agüera, A.; Fernández-Alba, A.R. Comparative study of analytical methods involving gas chromatography–mass spectrometry after derivatization and gas chromatography–tandem mass spectrometry for the determination of selected endocrine disrupting compounds in wastewaters. *J. Chromatogr. A* **2004**, *1047*, 129–135. [CrossRef]

9. Vega-Morales, T.; Sosa-Ferrera, Z.; Santana-Rodríguez, J.J. Determination of alkylphenol polyethoxylates, bisphenol A, 17α-ethynylestradiol and 17β-estradiol and its metabolites in sewage samples by SPE and LC/MS/MS. *J. Hazard. Mater.* **2010**, *183*, 701–711. [CrossRef]

10. Yiantzi, E.; Psillakis, E.; Tyrovola, K.; Kalogerakis, N. Vortex-assisted liquid–liquid microextraction of octylphenol, nonylphenol and bisphenol-A. *Talanta* **2010**, *80*, 2057–2062. [CrossRef]

11. Kapelewska, J.; Kotowska, U.; Karpińska, J.; Kowalczuk, D.; Arciszewska, A.; Świrydo, A. Occurrence, removal, mass loading and environmental risk assessment of emerging organic contaminants in leachates, groundwaters and wastewaters. *Microchem. J.* **2018**, *137*, 292–301. [CrossRef]

12. Grassi, M.; Rizzo, L.; Farina, A. Endocrine disruptors compounds, pharmaceuticals and personal care products in urban wastewater: Implications for agricultural reuse and their removal by adsorption process. *Environ. Sci. Pollut. Res.* **2013**, *20*, 3616–3628. [CrossRef] [PubMed]

13. Tsui, M.M.P.; Leung, H.W.; Lam, P.K.S.; Murphy, M.B. Seasonal occurrence, removal efficiencies and preliminary risk assessment of multiple classes of organic UV filters in wastewater treatment plants. *Water Res.* **2014**, *53*, 58–67. [CrossRef] [PubMed]

14. Lozano, N.; Rice, C.P.; Ramirez, M.; Torrents, A. Fate of triclocarban, triclosan and methyltriclosan during wastewater and biosolids treatment processes. *Water Res.* **2013**, *47*, 4519–4527. [CrossRef] [PubMed]

15. Basile, T.; Petrella, A.; Petrella, M.; Boghetich, G.; Petruzzelli, V.; Colasuonno, S.; Petruzzelli, D. Review of endocrine disrupting compound removal technologies in water and wastewater treatment plants: An EU perspective. *Ind. Eng. Chem. Res.* **2011**, *50*, 8389–8401. [CrossRef]

16. Koh, Y.K.K.; Chiu, T.Y.; Boobis, A.; Cartmell, E.; Scrimshaw, M.D.; Lester, J.N. Treatment and removal strategies for estrogens from wastewater. *Environ. Technol.* **2008**, *29*, 245–267. [CrossRef]

17. Tran, N.H.; Urase, T.; Ngo, H.H.; Hu, J.; Ong, S.L. Insight into metabolic and cometabolic activities of autotrophic and heterotrophic microorganisms in the biodegradation of emerging trace organic contaminants. *Bioresour. Technol.* **2013**, *146*, 721–731. [CrossRef]

18. Ternes, T.A.; Joss, A.; Siegrist, H. Scrutinizing pharmaceuticals and personal care products in wastewater treatment. *Environ. Sci. Technol.* **2004**, *38*, 392–399. [CrossRef]

19. Bartolomeu, M.; Neves, M.G.P.M.S.; Faustino, M.A.F.; Almeida, A. Wastewater chemical contaminants: Remediation by advanced oxidation processes. *Photochem. Photobiol. Sci.* **2018**, *17*, 1573–1594. [CrossRef]

20. Asgari, E.; Esrafili, A.; Rostami, R.; Farzadkia, M. O_3, O_3/UV and O_3/UV/ZnO for abatement of parabens in aqueous solutions: Effect of operational parameters and mineralization/biodegradability improvement. *Proc. Saf. Environ. Prot.* **2019**, *125*, 238–250. [CrossRef]

21. Luukkonen, T.; Teeriniemi, J.; Prokkola, H.; Rämö, J.; Lassi, U. Chemical aspects of peracetic acid based wastewater disinfection. *Water Res.* **2014**, *40*, 73–80.

22. Mielech-Łukasiewicz, K.; Starczewska, B. The use of boron-doped diamond electrode for the determination of selected biocides in water samples. *Water* **2019**, *11*, 1595. [CrossRef]

23. Hryniewicka, M.; Starczewska, B.; Gołębiewska, A. Determination of budesonide and sulfasalazine in water and wastewater samples using DLLME-SFO-HPLC-UV method. *Water* **2019**, *11*, 1581. [CrossRef]

24. Kotowska, U.; Kapelewska, J.; Kotowski, A.; Pietuszewska, E. Rapid and sensitive analysis of hormones and other emerging contaminants in groundwater using ultrasound-assisted emulsification microextraction with solidification of floating organic droplet followed by GC-MS Detection. *Water* **2019**, *11*, 1638. [CrossRef]
25. Ferrari, R.S.; de Souza, A.O.; Annunciação, D.L.R.; Sodré, F.F.; Dorta, D.J. Assessing surface sediment contamination by PBDE in a recharge point of guarani aquifer in Ribeirão Preto, Brazil. *Water* **2019**, *11*, 1601. [CrossRef]
26. Godlewska-Żyłkiewicz, B.; Sawicka, S. Removal of platinum and palladium from wastewater by means of biosorption on fungi *Aspergillus* sp. and yeast *Saccharomyces* sp. *Water* **2019**, *11*, 1522. [CrossRef]
27. Ambauen, N.N.; Mu, J.; Ngoc, N.L.; Hallé, C.; Trinh, T.T.; Meyn, T. Insights into the kinetics of intermediate formation during electrochemical oxidation of the organic model pollutant salicylic acid in chloride electrolyte. *Water* **2019**, *11*, 1322. [CrossRef]
28. Karpińska, J.; Sokol, A.; Koldys, J.; Ratkiewicz, A. Studies on the kinetics of doxazosin degradation in simulated environmental conditions and selected advanced oxidation processes. *Water* **2019**, *11*, 1001. [CrossRef]
29. Regulska, E.; Breczko, J.; Basa, A. Pristine and graphene-quantum-dots-decorated spinel nickel aluminate for water remediation from dyes and toxic pollutants. *Water* **2019**, *11*, 953. [CrossRef]
30. Mendoza, S.M.V.; Moreno, E.A.; Fajardo, C.A.G.; Medina, R.F. Liquid–liquid continuous extraction and fractional distillation for the removal of organic compounds from the wastewater of the oil industry. *Water* **2019**, *11*, 1452. [CrossRef]

water

MDPI

Article

The Use of Boron-Doped Diamond Electrode for the Determination of Selected Biocides in Water Samples

Katarzyna Mielech-Łukasiewicz *[ID] and Barbara Starczewska

Institute of Chemistry, University of Białystok, Ciołkowskiego 1 K, 15-245 Białystok, Poland
* Correspondence: mielech@uwb.edu.pl; Tel.: +48-85-738-80-65

Received: 24 June 2019; Accepted: 30 July 2019; Published: 31 July 2019

check for
updates

Abstract: In recent years, the remains of chemical substances in water environments, referred to as emerging organic contaminations, have been more and more often studied by analysts. This work shows the possibility of using a boron-doped diamond electrode to determine low concentration levels of remains of pharmaceuticals in environmental samples. The study focused on selected biocides from the group of azole fungicides (itraconazole and posaconazole) and was performed using quick and sensitive electrochemical methods. The cyclic voltammetry method was used in order to determine the properties of these compounds, whereas analytical characterization was performed using square wave voltammetry. The work involved the specification of the optimum electrooxidation conditions of the selected fungicides, their comparative characterization, and the development of a new, sensitive methods of itraconazole and posaconazole assay. The proposed procedures allowed us to determine itraconazole in the range from 7.9×10^{-8} to 1.2×10^{-6} mol·L^{-1} and posaconazole in the range from 5.7×10^{-8} to 8.44×10^{-7} mol·L^{-1}. The relative standard deviation of the measurements did not exceed 5.85%. The developed procedures were successfully used to determine itraconazole and posaconazole concentration in water samples and the assay recovery was between 93.5% and 102.8%.

Keywords: biocides; pollutants; water; boron-doped diamond electrode; electrochemical oxidation

1. Introduction

The issue of remains of pharmaceutical substances in water environments is becoming a global problem. The development of the pharmaceutical industry, the advancement of medicine, the increasing number of civilization diseases, the appearance of antibiotic-resistant bacteria and the growing consumption of drugs as a part of disease prevention lead to a dramatic growth in the amount of pharmaceutical contamination in water and wastewater. Particular attention is given to the remains of chemicals referred to as "emerging organic contaminants", such as active ingredients of pharmaceuticals, cosmetics, preservatives, and surfactants. In surface water, wastewater, and even drinking water, the remains of medicines and their active metabolites are more and more often found in amounts expressed in ng/L or µg/L. Water contaminated by them is a serious threat for the life and health of humans and animals. Classical methods of water purification are not effective in eliminating the broad spectrum of newly emerging pharmaceuticals, which get to drinking water, groundwater or bottom sediments. Therefore, studies related to the environmental impact assessment of pharmaceuticals are needed. It is necessary to develop analytical methods for different environmental matrices. The development of analytical methods for validation of pharmaceuticals in environmental matrices is becoming more and more important and necessary [1–6].

One group of compounds that have been given more and more attention recently is biocides [7]. Water contamination with these compounds results from their common use in daily life. They are used as active substances in pharmaceutical preparations or body care products (e.g., creams, ointments or shampoos) [8]. The presence of biocides has already been observed in sewage treatment plants and various environmental media [9–13]. For example, miconazole, ketoconazole and fluconazole have been detected in wastewater in concentrations up to 36, 90 and 140 ng/mL, respectively [14]. Itraconazole, fenticonazole and tioconazole have also been detected in sludge from a sewage treatment plant in real samples from the Northwest of Spain in concentrations of 204, 110 and 74 ng/g, respectively [15]. Pharmaceuticals in environmental samples are usually detected using gas or liquid chromatography methods, sometimes combined with tandem mass spectrometry [16]. Spectroscopic methods such as near infrared spectroscopy (NIR) or nuclear magnetic resonance (NMR) are also used [17].

Electrochemical methods and boron-doped diamond electrodes (BDD) are more and more often used to assay pharmaceuticals in environmental samples. The BDD is a new electrode material, effective also in the degradation and removal of contamination from water samples [18,19]. The BDD electrode allows us to measure the analyzed samples quickly, and the low and stable current ensures the high sensitivity of the measurements. Thanks to its unique physical and chemical properties [20–24], the boron-doped diamond electrode is an alternative to traditional carbon electrodes. The BDD electrode ensures excellent chemical resistance and stability in water environments. It has a very wide potential window, so it can be used in various electrochemical reactions in water environment. In addition, it has poor adsorption properties and high ability of oxidizing organic and inorganic compounds [25–30]. A conductive diamond has been used to assay biologically active compounds in various water matrices. This electrode has been used, in other words, to assay chemotherapeutics (e.g., ciprofloxacin) in natural waters and wastewater [31]. The BDD electrode has been used to assay various antibiotics (tetracycline, erythromycin, sulfamethoxazole), antidepressants (e.g., fluoxetine, viloxazine) and analgesics (e.g., naproxen) in environmental water samples [32–36].

In this work the use of the BDD electrode to assay selected biocides in water samples contaminated with pharmaceutical substances is proposed (Figure 1).

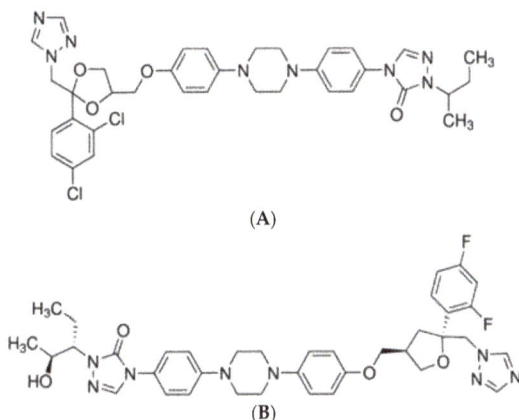

Figure 1. Structure of studied biocides: (**A**) itraconazole, (**B**) posaconazole.

2. Experimental

Apparatus and Chemicals

All the experiments were carried out using the Autolab PGSTAT 128N potentiostat from Metrohm Autolab B.V. with NOVA ver. 1.10 software, allowing for computer data collection and analysis of the results. The set used in voltammetric measurements included three electrodes. The working electrode was a D-035-SA boron-doped diamond electrode (BDD) from Windsor Scientific LTD, Slough, UK (diamond doped with boron, around 0.1%, A = 0.07 cm^2). The auxiliary electrode was a platinum plate with the surface area A = 0.9 cm^2. The reference electrode was EK-602 saturated calomel electrode (SCE) from Eurosensor.

Before each measurement series, the surface of BDD electrode was activated through cathodic reduction of hydrogen within the potential range of −2.9 V to 0.3 V. In the case of adsorption of the studied substance on the surface of the working electrode, it was mechanically purified using an MF-2060 polishing cloth from Bioanalytical System (USA) covered with aluminum oxide with grain sizes of 0.05 μm or 0.01 μm. All the experiments were carried out at room temperature and in the presence of oxygen. The studied solutions were not deoxidized before the voltammetric measurement.

In order to prepare the solutions, the analytes were weighted on a PB-153 scale (Mettler Toledo, Greifensee, Switzerland). The solutions' pH was measured with a pH-meter inoLab Level 1 from WTW, Germany. All the substances used in the experiments—itraconazole (ITR), posaconazole (POZ)—were purchased from Sigma-Aldrich, Hamburg, Germany. The solution of posaconazole with the concentration of 5.7×10^{-4} mol/L and the solution of itraconazole with the concentration of 1×10^{-3} mol/L were prepared by dissolving the appropriate weighted amounts of those substances in methanol. All the solutions were kept in a refrigerator.

The supporting electrolytes were 0.1 mol/L solution of potassium chloride, Britton-Robinson (B-R) buffer solutions with pH: 1.81; 2.29; 2.87; 4.35; 5.33; 7.0; 8.36; 9.15; 10.38; 11.20. B-R solutions with these pH values were prepared by mixing 0.4 mol/L H_3PO_4, 0.04 mol/L H_3BO_3 and 0.04 mol/L CH_3COOH with the appropriate amount of 0.2 mol/L NaOH.

The interferents were substances that may potentially occur in water, such as salts of sodium, potassium, iron(II), magnesium, copper, lead, calcium, cadmium and zinc, ions: Cl^-, $SO_4{}^{2-}$, $NO_3{}^-$, organic substances: Triton X-100, sodium dodecyl sulfate (SDS), tetrabutylamonium bromide, ketoconazole, clotrimazole, voriconazole, triclosan and methylparaben. The solutions of these salts with the concentrations of 2.8×10^{-4} mol/L were prepared by dissolving the appropriate weighted amounts of relevant substances in water. All the solutions were prepared using deionized water Milli–Q (Millipore Corp., Burlington, MA, USA).

3. Results and Discussion

3.1. Electrochemical Behavior of Selected Biocides

Azoles are the most popular and the most frequently used class of fungicides. First-generation azoles are imidazoles (e.g., ketoconazole), while second- and third-generation azoles are triazoles (e.g., posaconazole and itraconazole studied in the work). The selected azoles were studied using cyclic voltammetry (CV). In order to determine the optimum conditions of oxidation of selected analytes, the influence of the kind and pH of the used supporting electrolyte was studied. The experiments were carried out in Britton-Robinson buffer in the pH range of 1.8 to 11.2. Figure 2 presents the cyclic curves plotted in various buffer solutions for posaconazole.

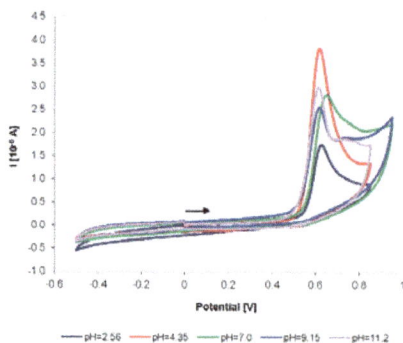

Figure 2. Cyclic voltammograms of 2.7×10^{-5} mol·L^{-1} posaconazole in Britton-Robinson (B-R) buffer in the pH 1.8–11.2 range; v = 100 mV·s^{-1}.

Analogically, oxidation curves of itraconazole as a compound structurally similar to posaconazole were plotted (Figure 3).

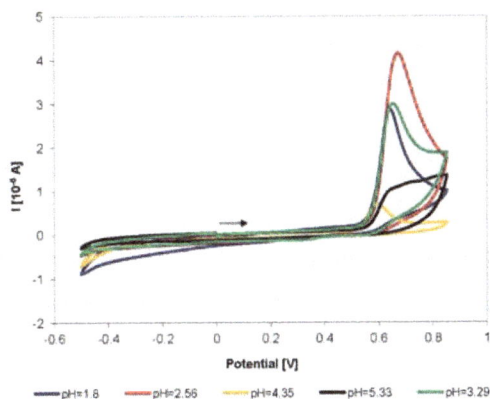

Figure 3. Cyclic voltammograms of 2.7×10^{-5} mol·L^{-1} itraconazole in B-R buffer in the pH 1.8–11.2 range; v = 100 mV·s^{-1}.

The voltammograms of all the studied compounds showed irreversible oxidation curves in the potential range of 0.58 V do 0.70 V. Figures 2 and 3 show example curves from all the studied period, and the complete characteristics of relationships between current and oxidation peak potential depending on pH are presented in Figure 4.

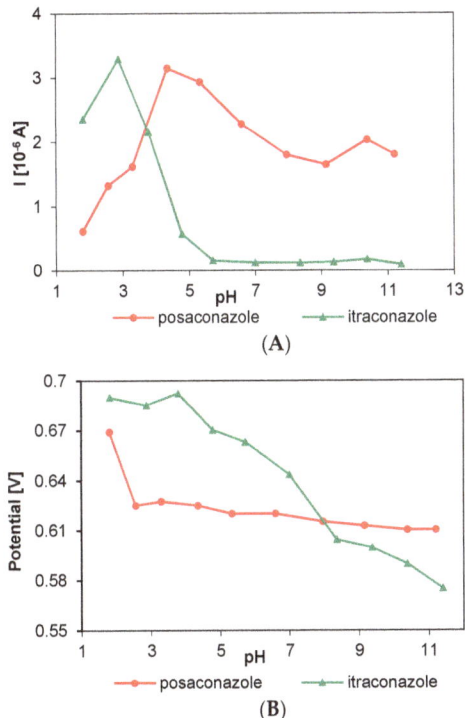

(A)

(B)

Figure 4. Effect of pH on selected biocides: (**A**) anodic peak current and (**B**) anodic peak potential.

The highest values of itraconazole oxidation peak were obtained in the B-R buffer with pH = 2.87, and of posaconazole in the solution with pH = 4.35. As the pH of the electrolyte increased, the potentials of recorded peaks of itraconazole and posaconazole moved towards lower values, and the current values decreased. The obtained results confirm literature data and earlier studies for ketoconazole [26,37]. The analysis of the above-mentioned azoles depending of the pH of the solution showed that the compounds' structure determines their electrochemical behavior. The comparable electrochemical behavior of these azole fungicides may prove the similarity of electrochemical reactions.

The most probable mechanism of oxidation of these compounds is the loss of an electron in the piperazine ring. Similar conclusions were also formulated for other substances containing a piperazine ring [38,39]. That oxidation produced lower currents with decreasing pH values. Considering the values: pKa, 3.6 for posaconazole and 3.7 for itraconazole [40,41], as well as the pH-dependence, the deprotonated piperazine group can be regarded as an electroactive form. This is also proved by the shift in the peak potential with the growth of pH. It shows that the deprotonated piperazine group must be produced by the dissociation of the proton [42].

3.2. Effect of Scan Rate

In order to determine the character of the recorded currents, the impact of the scan rate on the values of recorded currents was investigated. The experiments were carried out at the scan rates: 5, 10, 25, 50, 75, 100, 200, 350 and 500 mV/s in the range of potentials from −0.2 V to 0.95 V. The measurements were performed in Britton-Robinson buffer solutions with properly selected pH values at which the highest current values were recorded for different analytes.

The cyclic curves of itraconazole recorded at different scan rates are shown in Figure 5.

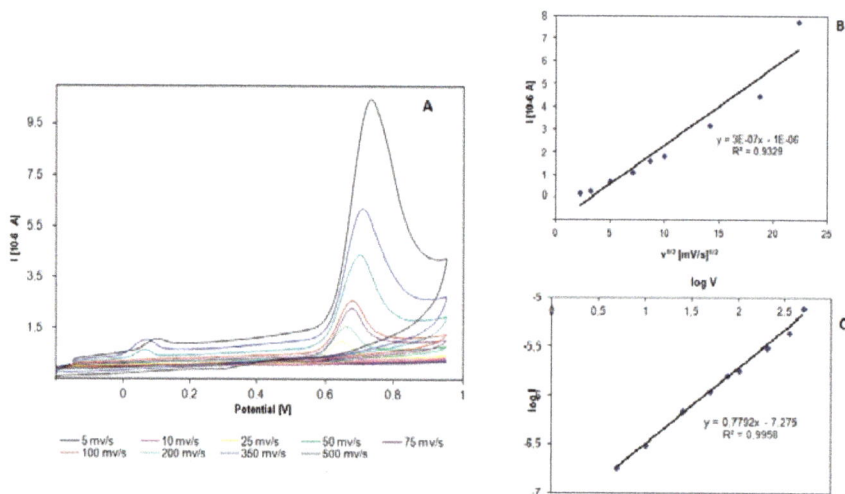

Figure 5. Cyclic voltammograms of 2.44×10^{-5} mol·L^{-1} itraconazole in B-R buffer pH 2.87 at different scan influence of interferents rates (**A**); the dependence of I_p on the square root of scan rate (**B**); and the plot for dependence of log I_p on log of scan rate (**C**).

The increase of the values of recorded currents along the increase of scan rate was observed for all the studied compounds. In the case of measurements for posaconazole, the analysis of peak current values of the studied compound depending on the root of the scan rate displayed a diffusion character. In the case of itraconazole, a clear decrease of the recorded current values was observed in successive cycles. This may be the evidence for itraconazole depositing on the surface of the electrode. The analysis of the logarithm of peak current of scan rate confirmed the diffusion character of currents in the case of posaconazole. The slope in the equations of curves was close to the theoretical value: 0.58 [43]. In the case of itraconazole, the slope of the curve was 0.77, which proves the mixed, diffusion and adsorption character of the measured currents. Figure 5C shows the relationship between the change in the logarithm of current and the logarithm of scan rate for the studied azole (itraconazole).

The equations describing these relationships can have the following form:

for itraconazole (ITR)

$$I\,(\mu A) = 3 \cdot 10^{-7}\,v^{1/2}\,(mVs^{-1})^{1/2} - 1 \cdot 10^{-6}$$

for posaconazole (POZ)

$$I\,(\mu A) = 9 \cdot 10^{-8}\,v^{1/2}\,(mVs^{-1})^{1/2} - 1 \cdot 10^{-7}$$

Based on the data from the obtained curves, the following equations were produced:

ITR

$$\log I_{pa}\,(\mu A) = 0.7792 \log v(mV/s) - 7275,\; R^2 = 0.9958$$

POZ

$$\log I_{pa}\,(\mu A) = 0.58 \log v(mV/s) - 7295,\; R^2 = 0.9996$$

3.3. Optimization of the Square Wave Parameters

The square wave voltammetry (SWV) technique was chosen to develop the procedure of assaying a compound from the group of azoles (itraconazole). It is one of the most sensitive electrochemical techniques, ensuring the high sensitivity of assays. The experimental parameters characteristic of SWV were determined as part of the work. The optimization of the procedure involved the amplitude, frequency and step potential. The amplitude was studied in the range of potentials from 5 to 100 mV. The measurements were performed at constant values of frequency and step potential. The increase in the amplitude value caused the increase in recorded currents, transition of peak potentials toward negative values, but also increased width of the peaks. Therefore, the optimum value of amplitude selected for further experiments is 25 mV. The impact of frequency on the height of recorded peaks was checked for the values: 8, 15, 25, 50, 75, 100, 125 and 150 Hz. The increase in the measured current was observed in the whole studied range. A linear relationship of current depending on the root of frequency was observed for the range of 8 Hz to 100 Hz. Hence, the value of 100 Hz was adopted as the best for further experiments. The study of the impact of step potential on the value of oxidation current of itraconazole showed that, as this parameter increased in the range of 2 mV to 10 mV, the recorded currents decreased. The value of 2 mV/s was chosen for further study. All the measurements were carried out in the B-R buffer solution with pH = 2.87 in the potential range of −0.2 V to 0.95 V.

The same procedure was carried out for posaconazole. The experiments led to the determination of the optimum amplitude value. Just like in the study of itraconazole, the best values were obtained for 25 mV. Testing frequencies between 8 and 150 Hz led to the choice of the optimum value of this parameter. The value of 100 Hz was chosen for the study, because as the frequency increased, the width of the recorded peaks also grew. The tests of step potential showed that the most symmetric peak was obtained for the value of 2 mV. The obtained values of parameters characteristic of the SWV method and the B-R buffer with pH = 4.35 were used in further quantitative study of posaconazole.

3.4. Analytical Curve for Itraconazole and Posaconazole Using SWV

The SWV technique was used to develop the procedure of assaying itraconazole and posaconazole. The study was carried out in experimentally confirmed optimum measurement conditions. In order to determine the relationship between the recorded current and concentration, currents were measured in solutions with growing concentrations of itraconazole. Four series of measurements were carried out, in the concentration range of 7.9×10^{-8} to 1.2×10^{-6} mol/L and with potentials from 0.2 V to 0.95 V.

The diagram of relationship between current values and the concentration of itraconazole on BDD electrode is presented in Figure 6.

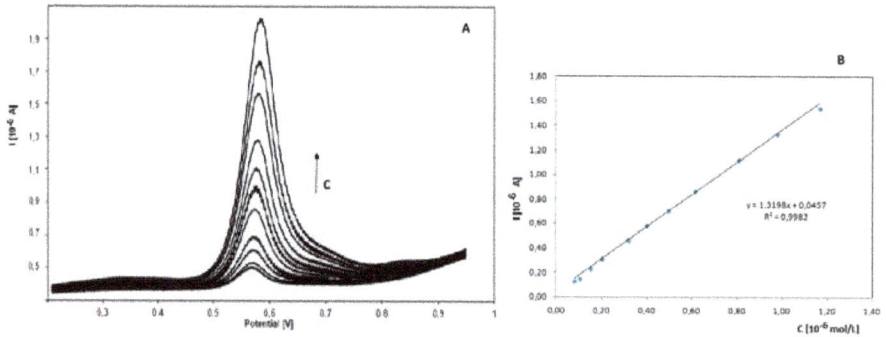

Figure 6. Square wave voltammetry (SWV) voltammograms of different concentrations of itraconazole in B-R buffer pH 2.87 (range of concentration 7.9×10^{-8} to 1.2×10^{-6} mol·L^{-1}) (**A**); Calibration curves of itraconazole using SWV methods, on the boron-doped diamond (BDD) electrode (**B**).

An analogous study of the relationship between the recorded current and the concentration was carried out for posaconazole. The measurements were carried out in the concentration range from 5.7×10^{-8} to 8.44×10^{-7} mol·L^{-1} and the potential range from 0.4 V to 0.75 V. The obtained calibration curve of posaconazole determination and an example series of obtained voltammograms are presented in Figure 7.

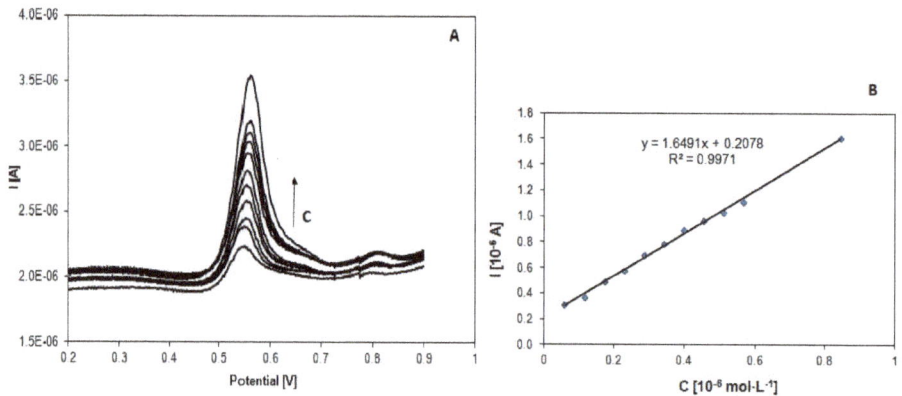

Figure 7. SWV voltammograms of different concentrations of posaconazole in B-R buffer pH 4.35 (range of concentration 5.7×10^{-8}–8.44×10^{-7} mol·L^{-1}) (**A**); Calibration curves of posaconazole using SWV methods, on the BDD electrode (**B**).

The obtained curves are well defined, sharp peaks. Increasing concentration results in higher current values, while the potential does not change. The obtained calibration curves have a coefficient of determination close to one, which proves good correlation of the results. The limit of detection (LOD) and limit of quantification (LOQ) were calculated on the basis of background curves obtained on BDD electrode using the SWV technique. The studied parameters were standard deviation from the obtained background currents (S) and the slope of the calibration curves (m). The following equations were used in the calculations [44]:

$$\text{LOD} = 3.3 \ S/m, \ \text{LOQ} = 10 \ S/m$$

Values characteristic of the calibration curves of itraconazole and posaconazole are presented in Table 1. The obtained data proves the high sensitivity of the proposed procedures.

The developed procedure determination of itraconazole was compared with other methods described in the literature (Table 2). The data presented in Table 2 proves the high sensitivity of the developed procedure against the methods proposed in the literature.

Table 1. Quantitative determination of itraconazole and posaconazole on the BDD electrode using the SWV method.

Studied Substance	Itraconazole	Posaconazole
Peak potential/V vs. SCE	0.59	0.55
Peak width half/mV	0.06	0.08
Linearity range/mol·L^{-1}	7.9×10^{-8}–1.2×10^{-6}	5.7×10^{-8}–8.44×10^{-7}
Slope/µA·L/mol	1.32×10^{6}	1.65×10^{6}
Intercept/µA	0.046	0.2
Correlation coefficient	0.9982	0.9971
LOQ/mol·L^{-1}	5.43×10^{-8}	2.36×10^{-8}
LOD/mol·L^{-1}	1.79×10^{-8}	7.78×10^{-9}
Repeatability of Ip/RSD%	5.68	1.73
Reproducibility of Ip/RSD%	2.61	1.93

Note: SCE: saturated calomel electrode; LOQ: limit of quantification; LOD: limit of detection.

Table 2. Comparison of linear range and detection limits for itraconazole to different methods.

Linear Range (mol·L^{-1})	Detection Limit (mol·L^{-1})	Method	Electrode	Ref.
2.2×10^{-8}–2.9×10^{-7}	1.9×10^{-8}		UTG	[45]
1.5×10^{-8}–2.3×10^{-7}	1.2×10^{-8}	AS-SWV	PG	
1.5×10^{-8}–1.5×10^{-7}	8.5×10^{-9}		CP	
2.2×10^{-8}–2.5×10^{-7}	1.5×10^{-8}		UTG	
4.5×10^{-8}–2.3×10^{-7}	1.2×10^{-8}	AS-DPV	PG	
1.5×10^{-8}–1.5×10^{-7}	1.1×10^{-8}		CP	
2.8×10^{-5}–1.4×10^{-4}	-	CV	GC	[46]
5.0×10^{-7}–5.0×10^{-6}	-	DPV	Hg	[47]
2.19×10^{-6}–6.33×10^{-5}	7.27×10^{-7}	CV	MWCNT/CP	[48]
7.9×10^{-8}–1.2×10^{-6}	1.79×10^{-8}	SWV	BDD	This work

Note: UTG ultra-trace graphite, PG pencil graphite, GC glassy carbon, CP carbon paste, BDD boron doped diamond, MWCNT multi-walled carbon nanotube, CV cyclic voltammetry, DPV differential pulse voltammetry, SWV square wave voltammetry, AS-SWV anodic stripping square wave voltammetry, AS-DPV anodic stripping differential pulse voltammetry.

The electrochemical procedure of posaconazole determination developed in the work is proposed in literature for the first time. It is a quick and simple procedure with high sensitivity, comparable with other methods, especially chromatographic ones [49].

3.5. Precision and Selectivity of Itraconazole Using SWV

The precision of the developed procedures was tested. The study of repeatability involved recording voltammetric curves of itraconazole with the concentration of 1.5×10^{-7} mol/L in the system of several repetitions for the same conditions of analysis. The measurements were performed within one day and over several successive days. The value of absolute standard deviation of measurements obtained within one day was 5.68%, and in the case of five successive days, 2.61%. Analogous studies for posaconazole solution with a concentration of 1.5×10^{-7} mol/L showed that the value of absolute standard deviation of measurements obtained within one day was 1.73%, and in the case of five successive days, 1.93%. The values of potentials of itraconazole or posaconazole oxidation peak did not change, and the obtained RSD values did not exceed 1%. The obtained results prove the good precision of the developed methods.

In order to study the selectivity of the developed methods, the impact of electrochemically active substances in the studied range of potentials and substances that can occur in environmental water samples was studied. The analysis involved interferents such as: Na^+, K^+, Cu^{2+}, Ca^{2+}, Fe^{2+}, Mg^{2+}, Cd^{2+}, Zn^{2+}, Pb^{2+}, Cl^-, SO_4^{2-}, NO_3^-, Triton X-100, sodium dodecyl sulfate (SDS), tetrabutylamonium bromide, and methylparaben. In addition, experiments were carried out for other fungicides (triclosan) and azole compounds (ketoconazole, clotrimazole, voriconazole). The tests were performed for itraconazole (or posaconazole) solution with the concentration 1.5×10^{-7} mol/L. Then, 10, 50, 100, 200 and 500-fold excess of the analyzed interferent in relation to the analyte was introduced. The heights of peaks on curves obtained in the presence of potentially interfering substances were compared with the height of peaks obtained for studied biocide alone. It was assumed that the compound does not cause interference if the approximation error of the assay does not exceed ±5%. Based on the obtained results, it was found that itraconazole can be assayed in the presence of inorganic ions with 500-fold excess of the interferent in relation to the concentration of the analyzed compound. Only in the case of Pb^{2+} ions, it was 100-fold excess in relation to the assayed analyte. Organic compounds such as: Triton X-100 (non-ionic surfactant), SDS (anionic surfactant), tetrabutylamonium bromide (cationic surfactant) and methylparaben allow to assay itraconazole at 100-fold excess of the interferent in relation to the assayed compound. The study of selectivity of itraconazole did not show the presence of new peaks on SWV curves in the presence of interferents. The observed influence was only the increase or decrease in the recorded peaks of the analyte in the presence of interferents (Figure 8).

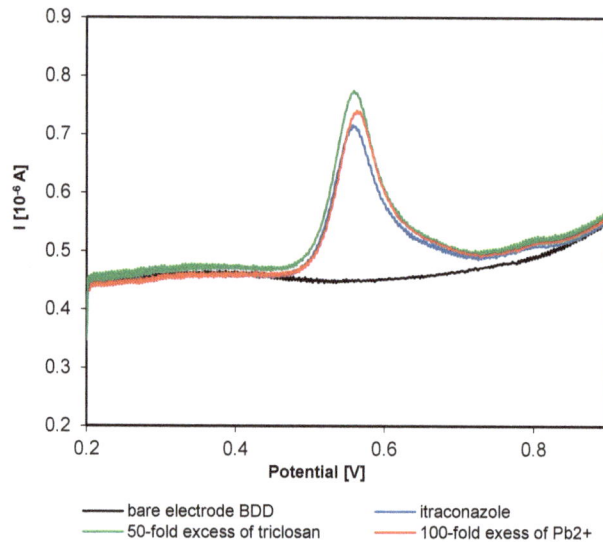

Figure 8. SWV voltammograms of itraconazole in B-R buffer pH 2.87 in the presents of selected interferents.

In the case of posaconazole, the recorded peaks increased at the 100-fold excess of Fe^{2+} and Cu^{2+} ions. The other ions did not change the peak heights in the studied range. Just like in the case of itraconazole, the study of surfactants showed interference at an over 100-fold excess, causing the increase in posaconazole oxidation peak. The measurements performed for other fungicides, such as triclosan and ketoconazole, showed the increase in the height of itraconazole and posaconazole oxidation peaks at a 5-fold excess. The study showed that triclosan and ketoconazole have the highest oxidation peaks in solutions with pH = 9. Both posaconazole and itraconazole can be assayed in the presence of azoles such as clotrimazole and voriconazole. Due to the structure of those azoles, different from that of itraconazole or posaconazole, oxidation peaks of these azoles are not observed in the studied potential range. The obtained results prove that the proposed methods display good selectivity.

3.6. Assay of Itraconazole and Posaconazole in the Water Samples

On the basis of the review of methods of assaying itraconazole described in the literature, we can say that biological liquids [45], pharmaceuticals [46] and water samples [11,15] are used as matrices to assay azole fungicides. So as to assess the practical utility of the developed procedure using square wave voltammetry (SWV) and BDD electrode, measurements were carried out in water matrices. The measurements were performed for samples of tap water and river water, which were enriched with known amounts of itraconazole. Methanol solution of itraconazole was added to tap or river water samples to obtain final concentrations: 1.5×10^{-7} mol/L, 3×10^{-7} mol/L and 4.5×10^{-7} mol/L, respectively. Three repetitions were performed for each prepared sample enriched with a specific amount of itraconazole, and then, the mean value was calculated.

The sample of river water was collected from the Biała river in Białystok, filtrated and kept in a refrigerator before the analysis. The sample of tap water was collected from the waterworks and analyzed without previous preparation. In order to assay itraconazole with the developed procedure in a real sample, the solutions were prepared by adding 10 mL water enriched with itraconazole to 10 mL B-R buffer with pH = 2.87. SWV curves were recorded in the range of potentials from 0.2 V to 0.9 V using previously optimized conditions (amplitude 0.025 V, frequency 100 Hz, step potential 2 mV). Figure 9 shows an example SWV curve of itraconazole assay in a real water sample.

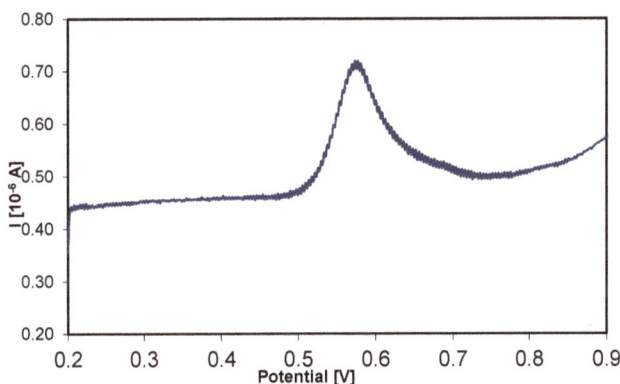

Figure 9. Square wave voltammogram of itraconazole in water samples.

Table 3 shows the results obtained for the assay of itraconazole in water samples. The experiments showed that voltammetric curves obtained for samples of water enriched with itraconazole did not differ in the shape or peak potential values from the peak recorded for the studied analyte in previous stages of the study. The rate of recovery in water samples ranged from 93.5% to 101%, which proves the good precision and accuracy of the developed method.

Table 3. Results of itraconazole determination in spiked water samples.

Sample	Amount Added/$\times 10^{-7}$ mol/L	Amount Received [a]/$\times 10^{-7}$ mol/L	Recovery [a]/%	RSD/%
Biała water	1.500	1.437 ± 0.069	95.8 ± 4.5	4.8
	3.000	2.882 ± 0.107	96.1 ± 3.6	3.7
	4.500	4.545 ± 0.102	101 ± 2.2	2.2
Tap water	1.500	1.403 ± 0.059	93.5 ± 3.9	4.2
	3.000	2.919 ± 0.056	97.3 ± 1.9	1.9
	4.500	4.350 ± 0.145	96.7 ± 3.2	3.3

[a] Mean value (n = 3).

The analysis of samples of tap water and river water enriched with known amounts of posaconazole was performed in the same way. In order to assay posaconazole in water samples using the developed procedure, we studied solutions with concentrations: 1.5×10^{-7} mol/L, 3×10^{-7} mol/L and 4.5×10^{-7} mol/L. SWV curves were recorded with the use of optimized apparatus parameters and B-R buffer with pH = 4.35. The results of the assay are included in Table 4. The rate of recovery from the measurements in water samples was between 94% and 102.8%. The obtained results confirm the analytical usability of the developed procedures of assaying itraconazole and posaconazole in samples of environmental water.

Table 4. Results of posaconazole determination in spiked water samples.

Sample	Amount Added/$\times 10^{-7}$ mol/L	Amount Received [a]/$\times 10^{-7}$ mol/L	Recovery [a]/%	RSD/%
Biała water	1.500	1.438 ± 0.052	95.8 ± 3.4	3.6
	3.000	2.821 ± 0.079	94.0 ± 2.6	2.8
	4.500	4.399 ± 0.115	97.7 ± 2.5	2.6
Tap water	1.500	1.447 ± 0.023	96.5 ± 1.6	1.6
	3.000	2.886 ± 0.102	96.2 ± 3.4	3.5
	4.500	4.627 ± 0.055	102.8 ± 1.2	1.2

[a] Mean value (n = 3).

4. Conclusions

The presented work shows the possibilities of using a new electrochemical electrode material—boron-doped diamond electrode—to assay compounds used as biocides (itraconazole and posaconazole). The electrochemical analysis showed that the selected compounds are electrochemically active and can well be studied and assayed on a BDD electrode. It is so because they undergo the reaction of irreversible electrochemical oxidation, and recorded peaks in the range of potentials from 0.55 V to 0.6 V are well defined and can be used in quantitative analysis. Experiments performed for itraconazole and posaconazole allowed us to develop a fully validated procedures, whose usability were verified for environmental water samples. The developed procedure had good sensitivity in comparison with other electrochemical methods of assaying itraconazole recommended in the literature [46,47]. Only the use of additional adsorption of the analyte on the electrode surface lowers the range of assays, but it makes the analysis longer and more complex [45]. The simplicity, speed and the lack of modification stages of BDD electrode are significant advantages of the proposed method. The electrochemical procedure of posaconazole determination developed in the work is proposed in the literature for the first time. It is a quick and simple procedure with high sensitivity, comparable with other methods, especially chromatographic ones [49]. Whereas, good rate of recovery (93.5–102.8%) proves that the developed procedures can be successfully used to assay itraconazole and posaconazole in water samples.

Author Contributions: Conceptualization, K.M.-Ł.; methodology, K.M.-Ł.; investigation, K.M.-Ł.; writing—original draft preparation, K.M.-Ł., writing—review and editing, K.M.-Ł. and B.S.; supervision, B.S.

Funding: This research received no external funding.

Conflicts of Interest: The authors declare no conflict of interest.

References

1. Gros, M.; Petrović, M.; Barcelo, D. Development of multi-residue analytical methodology based on liquid chromatography-tandem mass spectrometry (LC-MS/MS) for screening and trace level determination of pharmaceuticals in surface and wastewaters. *Talanta* **2006**, *70*, 678–690. [CrossRef]
2. Zounkova, R.; Kovalova, L.; Blaha, L.; Dott, W. Ecotoxicity and genotoxicity assessment of cytotoxic antineoplastic drugs and their metabolites. *Chemosphere* **2010**, *81*, 253–260. [CrossRef]

3. Stuart, M.; Lapworth, D.; Crane, E.; Hart, A. Review of risk from potential emerging contaminants in UK groundwater. *Sci. Total Environ.* **2012**, *416*, 1–21. [CrossRef]

4. Hernando, M.D.; Gomez, M.J.; Aguera, A.; Fernandez-Alba, A.R. LC-MS analysis of basic pharmaceuticals (beta-blockers and anti-ulcer agent) in wastewater and surface water. *Trends Anal. Chem.* **2007**, *26*, 581–594. [CrossRef]

5. Hyungkeun, R.; Subramanya, N.; Zhao, F.; Yu, C.; Sandt, J.; Chu, K. Biodegradation potential of wastewater micropollutants by ammonia-oxidizing bacteria. *Chemosphere* **2009**, *77*, 1087–1089.

6. Sirés, I.; Brillas, E. Remediation of water pollution caused by pharmaceutical residues based on electrochemical separation and degradation technologies: A review. *Environ. Int.* **2012**, *40*, 212–229. [CrossRef] [PubMed]

7. European Commission. Directive 98/8/EC of the European Parliament and of the Council (Biocidal Products Directive (BPD) 98/8/EC). *Off. J. Eur. Comm.* **1998**, *41*, 123.

8. Bester, K.; Scholes, L.; Wahlberg, C.; McArdell, C.S. Sources and mass flows of xenobiotics in urban water cycles—An overview on current knowledge and data gaps water. *Air Soil Pollut. Focus* **2008**, *8*, 407–4233. [CrossRef]

9. Chen, Z.; Ying, G.; Lai, H.; Chen, F.; Su, H.; Liu, Y.; Peng, F.; Zhao, J. Determination of bocides in differential environmental matrices by use of ultra-high-performance liquid chromatography-tandem mass spectrometry. *Anal. Bioanal. Chem.* **2012**, *404*, 3175–3188. [CrossRef]

10. Van De Steene, J.C.; Lambert, W.E. Validation of solid-phase extraction and liquid chromatography-electrospray tandem mass spectrometric method for the determination of nine basic pharmaceuticals in wastewater and surface water samples. *J. Chromatogr. A* **2008**, *1182*, 153–160. [CrossRef] [PubMed]

11. Van De Steene, J.C.; Lambert, W.E. Comparison of matrix effects in HPLC-MS/MS and UHPLC-MS/MS analysis of nine basic pharmaceuticals in surface waters. *J. Am. Soc. Mass Spectrom.* **2008**, *19*, 713–718. [CrossRef]

12. Wang, Z.; Zhao, P.; Yu, J.; Jiang, Z.; Guo, X. Experimental and molecular docking study on grapheme/Fe$_3$O$_4$ composites as a sorbent for magnetic solid-phase extraction of seven imidazole antifungals in environmental water samples prior to LC-MS/MS for enantiomeric analysis. *Microchem. J.* **2018**, *140*, 222–231. [CrossRef]

13. Lindberg, R.H.; Fick, J.; Tysklind, M. Screening of antimycotics in Swedish sewage treatment plants—Waters and sludge. *Water Res.* **2010**, *44*, 649–657. [CrossRef]

14. Van De Steene, J.C.; Stove, C.P.; Lambert, W.E. A field study on 8 pharmaceuticals and 1 pesticide in Belgium: Removal rates in waste water treatment plants and occurrence in surface water. *Sci. Total Environ.* **2010**, *408*, 3448–3453. [CrossRef] [PubMed]

15. Castro, G.; Roca, M.; Redriguez, I.; Ramil, M.; Cela, R. Identification and determination of chlorinated azoles in sludge using liquid chromatography quadrupole time-of-flight and triple quadrupole mass spectrometry platforms. *J. Chromatogr. A* **2016**, *1476*, 69–76. [CrossRef] [PubMed]

16. Fatta, D.; Nikolaou, A.; Achilleous, A.; Meric, S. Analytical methods for tracing pharmaceutical residues in water and wastewater. *Trends Anal. Chem.* **2007**, *26*, 515–535. [CrossRef]

17. Renitaa, A.; Senthil, P.; Kumarb, P.; Srinivasb, S.; Priyadharshinib, S.; Karthikab, M. A review on analytical methods and treatment techniques of pharmaceutical wastewater. *Desalin. Water Treat.* **2017**, *87*, 160–178. [CrossRef]

18. Peralta, E.; Natividad, R.; Roa, G.; Marin, R.; Romero, R.; Pavon, T. A comparative study on the electrochemical production of H$_2$O$_2$ between BDD and graphite cathodes. *Sustain. Environ. Res.* **2013**, *23*, 259–266.

19. Panizza, M.; Brillas, E.; Comninellis, C. Application of boron-doped diamond electrodes for wastewater treatment. *J. Environ. Eng. Manag.* **2008**, *18*, 139–153.

20. Einaga, Y. Diamond electrodes for electrochemical analysis. *J. Appl. Electrochem.* **2010**, *40*, 1807–1816. [CrossRef]

21. Luong, J.; Male, K.; Glennon, J. Boron-doped diamond electrode: Synthesis, characterization, functionalization and analytical applications. *Analyst* **2009**, *134*, 1965–1979. [CrossRef] [PubMed]

22. Zhou, Y.; Zhi, J. The application of boron-doped diamond electrode in amperometric biosensors. *Talanta* **2009**, *79*, 1189–1196. [CrossRef] [PubMed]

23. Yang, N.; Yu, S.; Macpherson, J.V.; Einaga, Y.; Zhao, H.; Zhao, G.; Swain, G.M.; Jiang, Y. Conductive diamond: Synthesis, properties, and electrochemical applications. *Chem. Soc. Rev.* **2019**, *48*, 157–204. [CrossRef] [PubMed]

24. Muzyka, K.; Sun, J.; Fereja, T.; Lan, Y.; Zhang, W.; Xu, G. Boron-doped diamond: Current progress and challenges in view of electroanalytical applications. *Anal. Methods* **2019**, *11*, 397–414. [CrossRef]
25. Pecková, K.; Musilová, J.; Barek, J. Boron-doped diamond film electrode—New tool for voltammetric determination of organic substances. *Crit. Rev. Anal. Chem.* **2009**, *39*, 148–172. [CrossRef]
26. Mielech-Łukasiewicz, K.; Rogińska, K. Voltammetric determination of antifungal agents in pharmaceuticals and cosmetics using a boron-doped diamond electrode. *Anal. Methods* **2014**, *6*, 7912–7922. [CrossRef]
27. Mielech-Łukasiewicz, K.; Dąbrowska, A. Comparison of boron-doped diamond and glassy carbon electrodes for determination of terbinafiny in pharmaceuticals using differential pulse and square wave voltammetry. *Anal. Lett.* **2014**, *74*, 1697–1711. [CrossRef]
28. Yardim, Y.; Alpar, N.; Senturk, Z. Voltammetric sensing of triclosan in the presence of cetyltrimethylammonium bromide using a cathodically pretreated boron-doped diamond electrode. *Int. J. Environ. Anal. Chem.* **2018**, *98*, 1–16. [CrossRef]
29. Brocenschi, R.; Rocha-Filho, R.; Biaggio, S.; Bocchi, N. DPV and SWV determination of estrone using a cathodically pretreated boron-doped diamond electrode. *Electroanalysis* **2014**, *26*, 1588–1597. [CrossRef]
30. Sousa, C.P.; Ribeiro, F.W.P.; Oliveira, T.M.; Salazar-Banda, G.R.; de Lima-Neto, P.; Morais, S.; Correia, A.N. Electroanalysis of pharmaceuticals on boron-doped diamond Electrodes: A Review. *ChemElectroChem* **2019**, *6*, 2350–2378.
31. Gayen, P.; Chaplin, B.P. Selective electrochemical detection of ciprofloxacin with a porous nafion/multi-walled carbon nanotube composite film electrode. *ACS Appl. Mater. Interfaces* **2016**, *8*, 1615–1626. [CrossRef] [PubMed]
32. Zhao, Y.; Yuan, F.; Quan, X.; Yu, H.; Chen, S.; Zhao, H.; Liub, Z.; Hilalb, N. An electrochemical sensor for selective determination of sulfamethoxazole in surface water using a molecularly imprinted polymer modified BDD electrode. *Anal. Methods* **2015**, *7*, 2693–2698. [CrossRef]
33. Radicova, M.; Behul, M.; Vojs, M.; Bodor, R.; Vojs Stano, A. Voltammetric determination of erythromycin in water samples using a boron-doped diamond electrode. *Phys. Status Solidi B* **2015**, *252*, 2608–2613. [CrossRef]
34. Calisto, C.; Cervini, P.; Cavalheiro, E.T. Determination of tetracycline in environmental water samples at a graphite-polyurethane composite electrode. *J. Braz. Chem. Soc.* **2012**, *23*, 938–943. [CrossRef]
35. Madej, M.; Kochana, J.; Baś, B. Determination of viloxazine by differential pulse voltammetry with boron-doped diamond electrode. *Monatshefte Chem.-Chem. Mon.* **2019**. [CrossRef]
36. Ardelean, M.; Manea, F.; Pop, A.; Schoonman, J. Carbon-based electrochemical detection of pharmaceuticals from water. *Int. J. Environ. Ecol. Eng.* **2016**, *10*, 1237–1242.
37. Vojić, M.; Popović, P. Protolytic equilibria in homogeneous and heterogeneous systems of ketoconazole and its direct spectrophotometric determination in tablets. *J. Serb. Chem. Soc.* **2005**, *70*, 67–78. [CrossRef]
38. Popa, O.M.; Diculescu, V.C. Electrochemical and spectrophotometric characterization of proteinkinase inhibitor and anticancer drug danusertib. *Electrochim. Acta* **2013**, *112*, 486–492. [CrossRef]
39. Uslu, B.; Topal, N.; Ozkan, S.A. Electroanalytical investigation and determination of pefloxacin in pharmaceuticals and serum at boron-doped diamond electrodes. *Talanta* **2008**, *74*, 1191–1200. [CrossRef] [PubMed]
40. Sanli, S.; Basaran, F.; Sanli, N.; Akmese, B.; Bulduk, A. Determination of dissociation constants of some antifungal drugs by two different methods at 298K. *J. Solut. Chem.* **2013**, 1976–1987. [CrossRef]
41. Courney, R.; Wexler, D.; Radwanski, E.; Lim, J.; Laughlin, M. Effect of food on the relative bioavailability of two oral formulations o posaconazole in healty adults. *Br. J. Clin. Pharmacol.* **2004**, *57*, 218–222. [CrossRef] [PubMed]
42. Alshalalfeh, M.; Sohail, M.; Saleh, T.; Aziz, A. Electrochemical investigation of gold nanoparticle-modified glassy carbon electrode and its application in ketoconazole determination. *Aust. J. Chem.* **2016**, *69*, 1314–1320. [CrossRef]
43. Galus, Z. *Fundamentals of Electrochemical Analysis*; Ellis Horwood Press: New York, NY, USA, 1994.
44. Miller, J.C.; Miller, J.N. *Statistics for Analytical Chemistry*; Ellis Horwood: Chichester, UK; New York, NY, USA, 1988.
45. Shalaby, A.; Hassan, W.S.; Hendawy, H.A.M.; Ibrahim, A.M. Electrochemical oxidation behavior of itraconazole at different electrodes and its anodic stripping determination in pharmaceuticals and biological fluids. *J. Electroanal. Chem.* **2016**, *763*, 51–62. [CrossRef]

46. Knoth, H.; Knoth, H.; Scriba, G.K.E.; Buettner, B. Electrochemical behavior of the antifungal agents itraconazole, posaconazole and ketoconazole at a glassy carbon electrode. *Pharmazie* **2015**, *70*, 374–378. [PubMed]

47. Knoth, H.; Petry, T.; Gartner, P. Differential pulse polarographic investigation of the antifungal drugs itraconazole, ketoconazole, fluconazole and voriconazole using a dropping mercury electrode. *Pharmazie* **2012**, *67*, 987–990.

48. Sultan, M.A.; Attia, A.K.; El-Alamin, M.M.A.; Atia, M.A. The novel use of multwalled carbon nanotubes-based sensors for voltammetric determination of itraconazole: Application to pharmaceutical dosage form and biological samples through spiked urine sample. *World J. Pharm. Sci.* **2016**, *5*, 93–108.

49. Reddy, T.M.; Tama, C.; Hayes, R.N. A dried blond spots technique based LC-MS/MS method for the analysis of posaconazole in human whole blood samples. *J. Chromatogr. B* **2011**, *879*, 3626–3638. [CrossRef]

water

MDPI

Article

Determination of Budesonide and Sulfasalazine in Water and Wastewater Samples Using DLLME-SFO-HPLC-UV Method

Marta Hryniewicka *, Barbara Starczewska and Agnieszka Gołębiewska

Institute of Chemistry, University of Białystok, ul. Ciołkowskiego 1K, 15-245 Białystok, Poland
* Correspondence: martah@uwb.edu.pl

Received: 18 June 2019; Accepted: 29 July 2019; Published: 30 July 2019

check for
updates

Abstract: Dispersive liquid–liquid microextraction based on solidification of floating organic droplet (DLLME-SFO) was applied to isolate budesonide (BUD) and sulfasalazine (SULF) from aqueous samples. The effects of different parameters on the efficiency on the extraction such as type of extrahent and dispersive solvent, ionic strength, pH of sample, and centrifugation time were investigated. Moreover, the influence of foreign substances on a studied process was tested. The calibration curves were recorded. The linearity ranges for BUD and SULF were 0.022–8.611 µg mL^{-1} and 0.020–7.968 µg mL^{-1} with the limit of detection (LOD) 0.011 µg mL^{-1} and 0.012 µg mL^{-1}, respectively. The enrichment factors (EF) for two analytes were high: for BUD it was 145.7 and for SULF, 119.5. The elaborated procedure was applied for HPLC-UV determination of these analytes in water and wastewater samples.

Keywords: budesonide; sulfasalazine; 1-undecanol; DLLME-SFO; HPLC-UV

1. Introduction

Increasing production and consumption of medicines contributes to the presence of biologically active substances in the environment and their concentration in surface waters. They affect negatively both the functioning of water reservoirs as well as living organisms despite the fact that pharmaceutical residues are present in very low concentration levels from ng L^{-1} to µg L^{-1}. Surface waters are important raw water sources for drinking water treatment plants, and a few studies have shown that pharmaceuticals and/or their metabolites may pass the treatment process and end up even in drinking water. In connection with the above, an important task for analysts is to monitor the concentrations of residues of biologically active compounds in surface waters and wastewater. Among the medicines often found in water and wastewater samples are: antibiotics, analgesics, beta-blockers, anti-inflammatory, and antidepressants [1].

Budesonide (BUD) belongs to glucocorticoids with a strong anti-inflammatory and antiallergic activity. It is used for the prevention and treatment of bronchial asthma and chronic obstructive pulmonary disease [2]. The controlled-release tablets are intended for people with intestinal inflammation or Crohn's disease of mild to moderate severity. Moreover, it is frequently used in veterinary medicine for the treatment of canine respiratory and bowel inflammatory diseases [3].

Sulfasalazine (SULF) is a synthetic drug and a combination of antibiotic (sulfapyridine) and an anti-inflammatory agent (5-aminosalysilic acid) which is extensively used in the treatment of inflammatory bowel diseases such as rheumatoid arthritis, Crohn's disease and ulcerative colitis [4].

There are several analytical methods available for the separate determination of BUD and SULF. BUD was determined in a variety of matrices using high-performance liquid chromatographic [5,6] and liquid chromatography–mass spectrometry (LC–MS or LC–MS/MS) [2,3,7–9]. Similarly, SULF was quantitated in a range of different matrices using high-performance liquid chromatography with diode array detection or ultraviolet detection (HPLC-DAD and HPLC-UV) [10–12], thin-layer chromatography densitometry [13], nuclear magnetic resonance (NMR) [14], spectroflourimetry [15] and liquid chromatography/positive-ion electrospray ionization mass spectrometry (LC-ESI(+)-MS/MS [16]. None of the above methods attempted the simultaneous analysis of BUD and SULF in the same sample.

The literature review shows that budesonide assays were made in biological samples [7–9], environmental samples [17,18], pharmaceutical formulations, and cosmetic products [19]. SULF was determined in pharmaceutical preparations [13–15], human plasma [10,16], water, and wastewater samples [20,21]. SULF was determined in river samples at the concentration of 15–76 ng L^{-1} and in wastewater samples at 65 ng L^{-1} (influent) and 266 ng L^{-1} (effluent) [20].

Traditionally, the sample treatment techniques used to isolate pharmaceuticals from water samples have been the liquid–liquid extraction (LLE) and the solid-phase extraction (SPE) [21]. LLE technique needs large volumes of toxic solvent, and the creation of emulsions is a common problem. However, SPE technique requires column conditioning and a process that is sometimes complicated and time-consuming. Therefore, the development of environmentally friendly pretreatment methods is necessary to overcome such disadvantages. Currently, microextraction techniques are often used to separate biologically active substances from aqueous solutions.

A novel dispersive liquid–liquid microextraction based on the solidification of floating organic drop (DLLME-SFO) was introduced by Leong et al. [22]. It involves the use of extraction solvent with a density lower than the density of water and freezing point near to the room temperature. The mixture of the dispersing and extracting solvent is injected into the water sample to form a turbid solution. After centrifugation, the tube is placed in an ice bath to solidify the extractant. The solidified drop is then collected and placed in a conical tube and allowed to melt. The liquid extract is analyzed by the appropriate instrumental technique. DLLME-SFO offers high analyte enrichment factors due to the use of small volumes of extractants compared to the volume of the water sample being tested. This method has gained recognition due to many advantages such as: simplicity of the procedure, high recovery, low cost, short extraction time, and the fact that it is not harmful to the environment.

On the ground of the review of literature, the use of the DLLME-SFO technique for the separation of organic [23–26], inorganic compounds and metal ions [27–30] was described in numerous publications presenting procedures for pretreatment of samples, e.g., tap water, water from the lake, human serum, urine, surficial sediments or beverage samples. A much smaller number of literature reports show the use of this technique for isolation of drugs. To the best of our knowledge, DLLME-SFO has not been applied for the isolation of BUD and SULF from water samples.

The aim of this study was to optimize the DLLME-SFO technique for isolating BUD and SULF from aqueous solutions. The effects of various experimental parameters such as a suitable extraction solvent and its volume, ionic strength, and centrifugation time were investigated. The simultaneous determination of these pharmaceuticals was performed by ultrahigh performance liquid chromatography-ultraviolet detection (HPLC-UV) method.

2. Experimental

2.1. Instrumentation

Thermo Separation chromatographic system with 2D Spectra System UV3000 (Panalytica, San Jose, CA, USA), a low-gradient pump P2000, a Rheodyne injector with 20-μL sample loop, and a vacuum membrane degasser SCM Thermo Separation were used (San Jose, California, CA, USA). The phase-separation process was accelerated by the centrifuge MPW-251 (MPW-Med.Instruments, Warsaw, Poland). A Hitachi U-1900 spectrophotometer (Panalytica, Tokyo, Japan) equipped with the deuterium discharge lamp and quartz cuvette was used for the measurements. Vibrating platform shaker (Heidolph, Vibramax 110, Schwabach, Germany) and magnetic stirrer hotplate (Heidolph MR3001K, Schwabach, Germany) were also applied.

2.2. Reagents and Standards

Budesonide (BUD, ≥99%, CAS 51333-22-3) and sulfasalazine (SULF, ≥98%, CAS 599-79-1) were bought from Sigma-Aldrich (Steinheim, Germany). Stock solutions of BUD or SULF containing 10^{-3} mol·L^{-1} of an analyte were prepared in methanol. Working solutions of these drugs were prepared freshly every day before analysis by diluting the standard solution with Milli-Q water and then it was stored in a dark bottle at room temperature.

Other chemicals at analytical grade used in experiments, like NaCl, KCl, CaCl$_2$, MgCl$_2$, KBr, Na$_2$SO$_4$, Fe(NO$_3$)$_3$, Na$_3$PO$_4$, and Na$_2$CO$_3$ (POCh, Gliwice, Poland), were prepared by dissolving an appropriate amount of salt in Milli-Q water. A working solution of HCl was prepared by successive dilutions of appropriate volumes of concentrated acid in Milli-Q water. Organic substances (studied interferences): diclofenac, ibuprofen, caffeine, acetylsalicylic acid, ascorbic acid, Levomepromazine, naproxen, and ranitidine were bought from Sigma Aldrich. Solvents such as acetonitrile and methanol at HPLC grade were purchased from J.T. Baker (USA).

2.3. Samples Preparation

The samples came from the Biała river, and the sewage was cleaned from the Municipal Sewage Treatment Plant in Białystok (Podlaskie Voivodeship, Poland). All samples were collected at the depth of 50 cm and were transferred to dark glass bottles. The total volume of each water and wastewater sample was 2 L. In the laboratory, samples were filtered through 0.22 µm pore size membrane filters to remove solid particles and analyzed within 24 h. The sample preparation was performed during the day in order to avoid the degradation of the analyte.

2.4. General DLLME-SFO Procedure

A mixture of methanolic solutions of BUD and SULF was introduced into 15 mL graduated tubes (the same concentration of both analytes). Subsequently, 50 µL of 3 mol L^{-1} HCl solution and 300 µL NaCl solution with a molar concentration of 1 mol L^{-1} were added. The tubes were supplemented with redistilled water to a volume of 10 mL. Mixtures of 1000 µL of ethanol (dispersant) and 100 µL of 1-undecanol (extractant) were prepared. In order to obtain a turbid solution, it was injected vigorously with a Pasteur pipette into the aqueous sample solution. The tubes were placed in a shaker (1500 rpm) for 10 min and then placed in a centrifuge to separate the organic phase from the aqueous and centrifuged for 15 min at 5000 rpm. The tubes were put into an ice bath for 10 minutes, the temperature of which was 3 °C. The solidified drop was collected and placed in another test tube for melting. The final volume of the extract was 70 ± 5 µL. The analysis was carried out using a high-performance liquid chromatograph with UV detection. Schematic diagram of DLLME-SFO extraction before HPLC-UV analysis was performed, as shown in Figure 1. The retention times for sulfasalazine and budesonide extracts are t_{R1} = 0.9 min (λ = 366 nm) and t_{R2} = 2.0 min (λ = 241 nm).

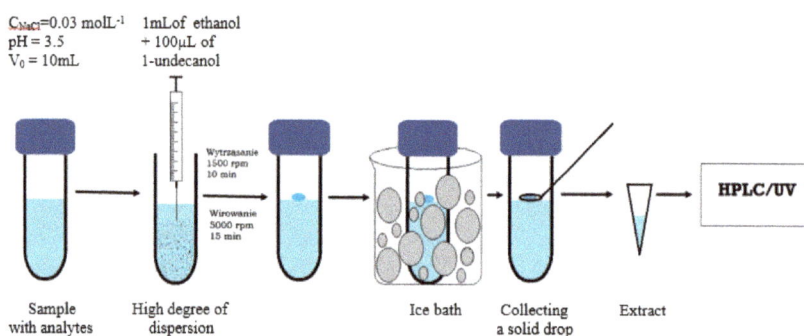

Figure 1. Schematic diagram of DLLME-SFO extraction before HPLC-UV analysis.

3. Results and Discussion

3.1. Primary Studies and HPLC Analysis

BUD (MW = 430.53 g mol^{-1}, log Kow = 2.4–2.7) and SULF (MW = 398.39 g mol^{-1}, log Kow = 3.7–4.8) are soluble substances in methanol (Figure 2). The absorption spectrum of the methanolic solution of these substances at a concentration of 5×10^{-5} mol L^{-1} were recorded. BUD has an absorption maximum at 241 nm and SULF at 366 nm (Figure 3).

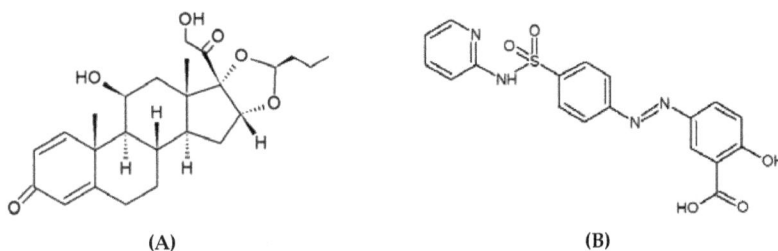

Figure 2. Molecule structure of: (**A**) budesonide (C$_{25}$H$_{34}$O$_6$, MW = 430.53 g mol^{-1}) and (**B**) sulfasalazine (C$_{18}$H$_{14}$N$_4$O$_5$S, MW = 398.39 g mol^{-1}).

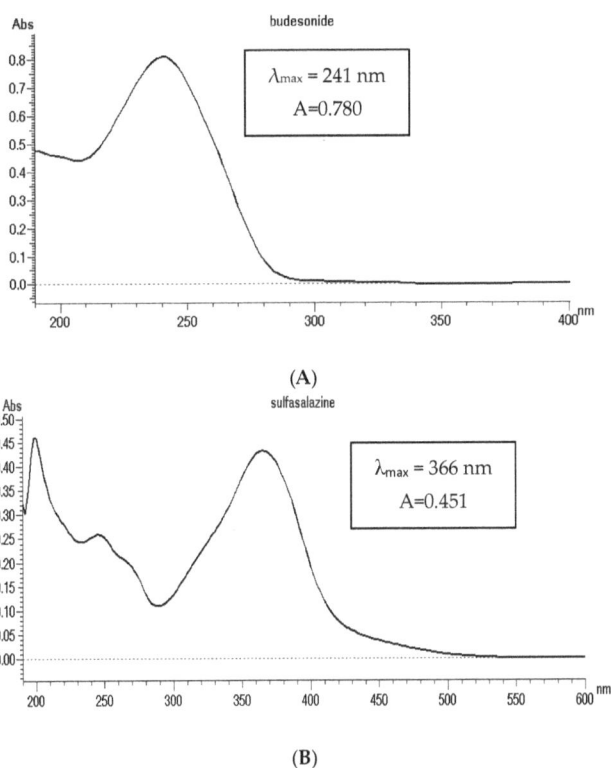

(**A**)

(**B**)

Figure 3. The absorption spectrum of budesonide (**A**) and sulfasalazine (**B**) solution in methanol (C = 5×10^{-5} mol L^{-1}).

The chromatographic separation was performed on a Lichrospher 100 RP-18 column (125 mm × 4.6 mm, 5 µm). In order to obtain optimal conditions for HPLC-UV chromatography analysis, a suitable mobile phase was chosen to ensure proper peaks in the chromatogram. The literature review shows that the following phases were used for the determination of BUD: acetonitrile–phosphate buffer (pH 3.2) (55:45, v/v) [5], methanol–water (80:20, v/v) [6], and for SULF: methanol–acetate buffer (48:52, v/v) [11], acetonitrile–acetate buffer (gradient elution) [12]. It was decided to check such phases as: methanol–water (70:30, v/v) and acetonitrile–water (50:50). The most symmetrical and highest peaks were obtained when the methanol–water phase was used. The selected mobile phase was also checked at various volumetric ratios of methanol and water. A flow rate of 1 mL min^{-1} was maintained. The injection volume was 50 µL. The typical chromatogram is presented in Figure 4.

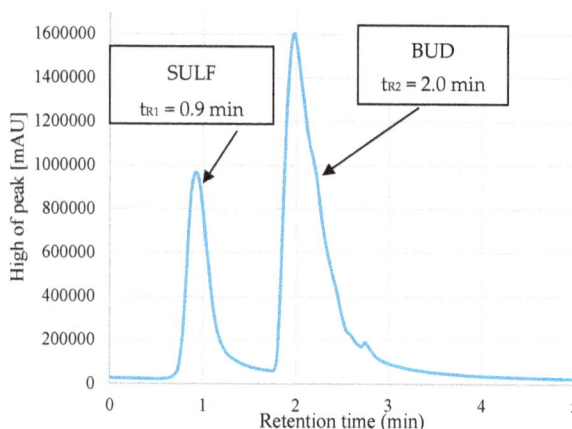

Figure 4. Chromatogram of extracts of BUD and SULF after DLLME-SFO.

3.2. Optimization of Extraction Parameters

The selection of appropriate DLLME-SFO extraction conditions is carried out in order to obtain the highest possible process efficiency. The following factors were optimized: solvent selection, electrolyte type and concentration, pH, and time and speed of shaking and centrifuging the sample.

3.2.1. Selection of Extraction Solvent and Its Volume

Optimizing solvents is critical for obtaining good extraction recoveries in DLLME-SFO method. The extraction solvent significantly influences the extraction efficiency. The selection of solvents for the isolation of BUD and SULF using the DLLME-SFO technique from aqueous solutions was checked. Solvents used for DLLME-SFO extraction have a density less than water and a freezing point close to room temperature (Table 1). The following solvents were used to isolate BUD and SULF: 1-dodecanol ($C_{12}H_{25}OH$) and 1-undecanol ($C_{11}H_{23}OH$). The density of these extractants is 0.83 g cm^{-3}, and the pour point is 22–24 °C (1-dodecanol) and 13–15 °C (1-undecanol). It has been found that the extraction of selected analytes can be used for both 1-dodecanol and 1-undecanol, however, the highest extraction efficiency for BUD and SULF were obtained by using 1-undecanol (Figure 5). Recovery for BUD was 70%, whereas for SULF it was 65%. It was decided to check its volume in order to obtain high recoveries. The volume of 1-undecanol in the range of 25–200 µL was tested. A series of solutions were prepared and injected with 1 mL of the dispersant (methanol) and variable volumes of the extractant (Figure 6). The highest efficiency of DLLME-SFO extraction for both analytes was obtained using 100 µL of 1-undecanol. Increasing the volume of the extractant resulted in a decrease in recovery and a simultaneous increase in the volume of the extract.

Table 1. Extraction solvents used in DLLME-SFO.

Extractant	Chemical Formula	Density (g cm^{-3})	Temperature of Solidification ($^\circ$C)
n-hexadecane	$CH_3(CH_2)_{14}CH_3$	0.77	18
2-dodecanol	$CH_3(CH_2)_9CH(OH)CH_3$	0.80	17–18
1-decanol	$CH_3(CH_2)_9OH$	0.83	6.4
1-dodecanol	$CH_3(CH_2)_{11}OH$	0.83	22–24
1-undecanol	$CH_3(CH_2)_{10}OH$	0.83	13–15
1-chlorooctadecane	$CH_3(CH_2)_{16}CH_2Cl$	0.85	20–24
1-bromohexadecane	$CH_3(CH_2)_{15}Br$	0.99	16–18
1,10-dichlorodecane	$Cl(CH_2)_{10}Cl$	0.99	14–16

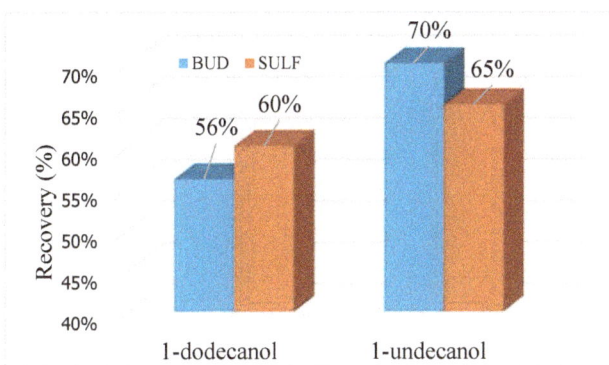

Figure 5. Recovery of analytes depending on the type of extractant.

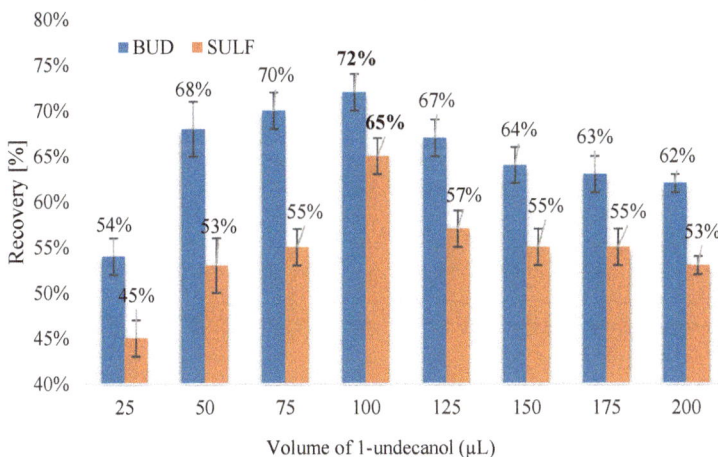

Figure 6. Recovery of analytes depending on the volume of 1-undecanol.

3.2.2. Selection of Dispersant Solvent and its Volume

Choosing the right dispersant is an important element in optimizing the extraction conditions. The solvent must dissolve in both the organic and aqueous phase and cause the formation of a high extractant dispersion when the extraction and dispersing solvent is injected into the water sample. It should also allow the analyte to enter the organic phase. The volume of the dispersing solvent used is usually in the range of 0.5 to 2.5 mL or more. This affects the degree of dispersion in the sample, and therefore the extraction efficiency. Too large disperser volumes may reduce the extraction efficiency. This is due to the increased solubility of analytes in water as the dispersant volume increases.

The effect of such dispersants as acetonitrile, methanol and ethanol was checked. The highest efficiency of extraction was obtained using ethanol. The optimal dispersant volume was also checked. A series of solutions were prepared into which were injected mixtures of varying volumes of the dispersing solvent from 250 to 1500 μL and 100 μL 1-undecanol (extraction solvent). The tubes were then placed in a shaker and centrifuged at 5000 rpm for 10 minutes. The samples were sequentially placed in an ice bath to solidify the extractant. The optimal volume of ethanol was determined to be 1 mL (Figure 7). Too small volume of the dispersant makes it impossible to obtain a high dispersion of the extractant in the entire volume of the water sample. The reported decrease in the extraction efficiency as the volume of ethanol increased is due to the increase in the affinity of the analyte for the aqueous phase.

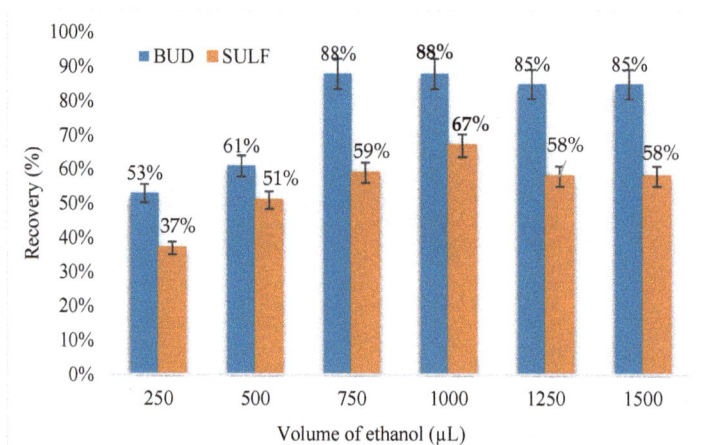

Figure 7. Recovery of analytes depending on the volume of ethanol (dispersing solvent).

3.2.3. Effect of Ionic Strength

The addition of salt solution increases the efficiency of extraction because the ionic strength of the solution is increased and the salting out effect occurs, which promotes the separation of the two phases and reduces the loss of the extractant. The effect of the addition of 300 μL electrolyte solutions at a concentration of 1 mol L^{-1} of the following salts was tested: sodium chloride (NaCl), potassium chloride (KCl) and calcium chloride (CaCl$_2$). The best extraction efficiency was observed using a sodium chloride solution (Table 2). After electrolyte selection, the concentration in the water sample was selected. A series of solutions were prepared to which variable volumes of NaCl solution with an initial concentration of 1 mol L^{-1} were added. It was found that the optimal concentration of NaCl for the extraction of both analytes is 3×10^{-2} mol L^{-1}. The effect of sodium chloride addition before and after injection of the dispersant and extractant mixture was also investigated. It has been found that the order of the electrolyte additive does not affect the efficiency of the extraction.

Table 2. Influence of the type of electrolyte on the efficiency of analyte extraction.

	Type of Electrolyte	Recovery (%)
BUD	NaCl	90
	KCl	57
	CaCl$_2$	72
SULF	NaCl	78
	KCl	64
	CaCl$_2$	73

3.2.4. Influence of pH of Sample

The choice of pH on the efficiency of DLLME-SFO of BUD and SULF was checked. An aqueous sample containing BUD has a pH in the range of 5–6, and the water sample containing SULF has a pH of about 2–3. This parameter was changed in the range 2–12 by the addition of an appropriate volume of solutions of 3 mol L^{-1} of hydrochloric acid and 2 mol L^{-1} sodium hydroxide. The microextraction was performed by adding to the aqueous samples the optimized volume of the dispersant (1 mL of ethanol) and extractant (100 μL 1-undecanol) mixture. It was found that the addition of the NaOH solution significantly reduces the efficiency of extraction of both analytes, while the acidic environment causes an increase in the recovery value. The greatest value is observed with the addition of 50 μL 3 mol L^{-1} hydrochloric acid solution. The final pH of the sample was 2.4.

3.2.5. Effect of the Time and Speed of Shaking and Centrifugation

Another parameter optimized during DLLME-SFO was the time and speed of shaking the sample. This process results in thorough mixing of the sample components and determining the equilibrium in the solution. Samples were shaken successively for 5 and 10 min at four speeds ranging from 750 to 1500 rpm. The extraction efficiency without the sample shaking step was also checked. After optimizing, the samples were shaken in the further studies at 1500 rpm for 10 min.

The next stage of the DLLME-SFO procedure was the centrifugation of the solution in order to separate the organic phase from the aqueous one. The effect of spin time and speed was changed by changing individual centrifugation programs. They were centrifuged successively for 5, 10 and 15 min at three centrifugation speeds. It was found that the optimal spinning speed for both analytes is 5000 rpm for 15 minutes. Using this centrifugation program enables the highest efficiency of DLLME-SFO extraction. Too short a centrifugation time contributes to inaccurate phase separation. The use of too high spin speeds results in a lower extraction efficiency.

3.3. Selectivity

The selectivity of the DLLME-SFO procedure against other biologically active substances was studied. These substances were: diclofenac, ibuprofen, metronidazole, caffeine, acetylsalicylic acid, ascorbic acid, levomepromazine, naproxen, and ranitidine. These compounds belong to different therapeutic classes, e.g., anti-inflammatory, analgesic or neuroleptic drugs. There is a probability of occurrence of these compounds in water and wastewater samples, which is why they have been selected as potential interferents for the determination of BUD and SULF. Moreover, the influence of foreign ions present in water and wastewater samples such as Mg^{2+}, Ca^{2+}, Fe^{3+}, SO$_4^{2-}$, PO$_4^{3-}$, and CO$_3^{2-}$ was investigated (Table 3). The tolerance limits were defined as the level of foreign substances causing an error of ±5% in determination of analytes. It was observed that DLLME-SFO procedure is tolerant to inorganic ions (acceptable excesses are usually in the range of 20–50) with the exception of iron ions. In the case of pharmaceuticals, the developed procedure is resistant to the presence of biological active compounds such as metronidazole or ranitidine. The literature review shows that selected chemical compounds have low octanol–water partition coefficients (log Kow), for caffeine it is

−0.07, for acetylsalicylic acid 1.25 and for metronidazole −0.02. In the DLLME-SFO extraction, low polarity solvents like 1-undecanol cannot extract polar compounds.

Table 3. The excess of interferences causes 5% error in HPLC-UV determination of BUD and SULF.

Interferent	BUD (5×10^{-5} mol L^{-1})	SULF (5×10^{-5} mol L^{-1})
Diclofenac	5	15
Ibuprofen	5	5
Metronidazole	10	15
Caffeine	10	10
Acetylsalicylic acid	20	3
Ascorbic acid	15	15
Levomepromazine	20	5
Naproxen	5	2
Ranitidine	20	10
Mg^{2+}	20	15
Ca^{2+}	20	20
Fe^{3+}	3	2
SO_4^{2-}	30	30
PO_4^{3-}	50	40
CO_3^{2-}	30	30

In summary, the developed method allows to determine BUD and SULF in the presence of other compounds, so it is highly selective for selected chemical compounds.

3.4. Analytical Performance

The standard curves for the determination of BUD and SULF before and after the extraction were recorded (five replicates). Chromatograms of extracts were recorded in which the concentration of analytes ranged from 0.022–8.611 µg mL^{-1} and 0.020–7.968 µg mL^{-1} for BUD and SULF, respectively. The calibration graphs for both analytes were linear over these concentrations. Correlation coefficients of curves equal to 0.999 ± 0.004 were obtained. Limit of detection (LOD) and quantification (LOQ) were calculated using standard deviation values (s) from 10 independent samples and the slope value from the calibration graphs (a). The formulas: LOD = 3.3 s/a and LOQ = 10 s/a were used. Under the optimum experimental conditions, the limit of detection (LOD) equal to 0.011 µg mL^{-1} for BUD and 0.012 µg mL^{-1} for SULF was achieved.

The enrichment factor (EF) was obtained from the ratio of the calibration curve slopes with and without the preconcentration step. EFs were found to be 145.7 for BUD and 119.5 for SULF. The final amount of 1-undecanol after the DLLME-SFO process is 70 ± 5µL. Microextraction techniques allow the use of small volumes of extractants. As a result, high enrichment factors are obtained, which allows to reduce LOD and LOQ. The extraction recovery (R%) can be calculated as follows: R (%) = V_{sed}/V_{aq} × EF × 100 where V_{sed} *and* V_{aq} are the volumes of the sedimented phase (0.07 mL) and sample solution (10 mL), respectively [31]. High recovery rates were also obtained: 102% ± 7% and 84% ± 5% for BUD and SULF, respectively.

The statistical evaluation of the recorded standard curves of BUD and SULF determination after DLLME-SFO extraction was performed. The coefficient of variation of the method for both analytes does not exceed 5%, which proves the accuracy of the analytical procedure developed. The precision of DLLME-SFO-HPLC-UV was evaluated over 10 independent replicates at a concentration of 1×10^{-5} mol L^{-1} during 1 day (intraday) and within 3 days (interday). The relative standard deviation

(RSD) of the measurements was less than 4%. The analytical characteristic data for the proposed method are summarized in Table 4.

Table 4. Analytical characteristic of DLLME-SFO-HPLC-UV method.

	BUD	SULF
Beer's low range (mol L^{-1})	5×10^{-8}–2×10^{-5}	5×10^{-8}–2×10^{-5}
Beer's low range (µg mL^{-1})	0.022–8.611	0.020–7.968
Equation of calibration graph (n = 5)	$y = 1.09 \times 10^{11}\,x + 6\,971$	$y = 1.15 \times 10^{11}\,x - 6\,390$
Slope ± standard deviation SD	$1.09 \times 10^{11} \pm 0.88 \times 10^{10}$	$1.15 \times 10^{11} \pm 0.98 \times 10^{10}$
Intercept ± standard deviation SD	$6\,971 \pm 1\,102$	$6\,390 \pm 989$
Correlation coefficient R^2 ± standard deviation SD	0.999 ± 0.004	
Precision—intraday RSD (n = 10, %)	3.75	3.15
Precision—interday RSD (n = 10, %)	0.66	2.88
Limit of detection LOD (mol L^{-1})	2.67×10^{-8}	2.92×10^{-8}
Limit of detection LOD (µg mL^{-1})	0.011	0.012
Limit of quantification LOQ (mol L^{-1})	8.10×10^{-8}	8.84×10^{-8}
Limit of quantification LOQ (µg mL^{-1})	0.035	0.035
Enrichment factor EF	145.7	119.5
Recovery ± standard deviation SD (%)	102 ± 7	84 ± 5
Volume of extract (µL)	70 ± 5	

In comparison with other techniques (Table 5), the developed method (DLLME-SFO-HPLC-UV) is characterized by high EFs values (in other methods, this parameter is skipped) and relatively low values of LOQ. Typically, the SPE technique is used to separate BUD and SULF from matrices using Oasis MCX or HLB sorbents. These sorbents can be repeatedly used, but the necessary step is the conditioning of the sorbent using relatively large amounts of solvents. The other microextraction techniques used to isolate these drugs are unknown. So, it is the first time that the microextraction technique (DLLME-SFO) which requires the low consumption of organic solvent is used for simultaneous isolation of BUD and SULF from environmental matrices. In addition, the LC-MS/MS method is a commonly used method that allows to achieve lower LOD and LOQ values, but requires the use of solvents of higher purity and consequently is more expensive in comparison with HPLC-UV.

Table 5. The comparison of elaborated procedure (DLLME-SFO-HPLC-UV) with other techniques.

Sample	Isolation Technique	Determination Method	LOQ	EF	Lit.
		BUD			
Surface water, wastewater	SPE	LC-MS/MS	4.2–5.8 ng L^{-1}	nd	[17]
Soils	SPE (Oasis HLB sorbent)	LC-MS/MS	2.84 ng g^{-1}	nd	[18]
Surface water, wastewater samples	DLLME-SFO	HPLC-UV	0.035 µg mL^{-1}	145.7	This method
		SULF			
Surface water	SPE	UPLC-MS/MS	5 ng L^{-1}	nd	[21]
Surface water, wastewater	SPE (Oasis MCX sorbent)	UPLC-ESI/MS/MS	1.5 ng L^{-1}	nd	[20]
Human serum	nd	HPLC-DAD	0.1 ng µL^{-1}	nd	[10]
Human serum	nd	HPLC-UV	0.5 µg mL^{-1}	nd	[32]
Surface water, wastewater	DLLME-SFO	HPLC-UV	0.035 µg mL^{-1}	119.5	This method

Nd - No data.

3.5. Application to Natural Samples

The method was successfully applied for the determination of SULF and BUD in natural samples. Two types of water (river water and wastewater samples) were analyzed by HPLC-UV after the DLLME-SFO procedure. The samples came from the Biała river, and the wastewater samples from the Municipal Sewage Treatment Plant (Poland).

The samples were then spiked with analytes standard solutions at different levels to assess matrix effects and the corresponding relative recoveries. Three samples of river water and wastewater were prepared, to which known amounts of budesonide and sulfasalazine were added (the concentration of added analytes was 5×10^{-6} mol L^{-1}). Subsequently, the DLLME-SFO extraction procedure and HPLC-UV measurement were carried out. The results are summarized in Table 6.

Table 6. Results of BUD and SULF in Biała river and wastewater samples by the proposed chromatographic method with isolation by DLLME-SFO extraction.

Sample	Added Concentration of Analyte (mol L^{-1})	Concentration of Found Analyte (mol L^{-1})	Average Concentration of Found Analyte ± SD (n = 3) (mol L^{-1})	RSD (n = 3, %)	Average Recovery ± SD (%)
		BUD			
Biała river	5.00×10^{-6}	4.68×10^{-6}	$5.11 \times 10^{-6} \pm 3.97 \times 10^{-7}$	7.8	102.1 ± 7.7
		5.16×10^{-6}			
		5.47×10^{-6}			
Wastewater		5.84×10^{-6}	$5.85 \times 10^{-6} \pm 1.02 \times 10^{-7}$	1.7	117.1 ± 1.7
		5.76×10^{-6}			
		5.96×10^{-6}			
		SULF			
Biała river	5.00×10^{-6}	4.37×10^{-6}	$4.78 \times 10^{-6} \pm 4.03 \times 10^{-7}$	8.4	95.5 ± 8.3
		5.17×10^{-6}			
		4.79×10^{-6}			
Wastewater		5.15×10^{-6}	$5.34 \times 10^{-6} \pm 3.06 \times 10^{-7}$	5.7	106.7 ± 5.7
		5.69×10^{-6}			
		5.17×10^{-6}			

Based on the results, it can be concluded that the developed chromatographic procedure for the determination of BUD and SULF combined with dispersive liquid–liquid microextraction based on solidification of floating organic droplet is an appropriate method. BUD and SULF are not found in river water and wastewater samples in the range of concentrations tested.

4. Conclusion

A simple and practical preconcentration technique, dispersive liquid–liquid microextraction based on the solidification of a floating organic drop (DLLME-SFO) combined with HPLC-UV has been evaluated for simultaneous extraction and determination of BUD and SULF. This is the first time that the DLLME-SFO technique was introduced for isolating these analytes from water and wastewater samples. The effective experimental parameters on the extraction efficiency such as extraction and dispersive solvents, ionic strength, and pH were studied. Under the optimum experimental conditions, the enrichment factors of 145.7 and 119.5 for BUD and SULF were obtained, respectively.

The calibration curves were linear for BUD and SULF in the range of 0.022–8.611 µg mL^{-1} and 0.020–7.968 µg mL^{-1} with the limit of detection (LOD) 0.011 µg mL^{-1} and 0.012 µg mL^{-1}, respectively. To the best of our knowledge, no method of determination of BUD and SULF described so far is characterized by such high EF values. This is the main advantage of the described method. Moreover, high sensitivity, good isolation of analytes in less than 40 min method time and very low solvent consumption (volume of extrahent equals 100 µL) were achieved. Additionally, a significant reduction in the usage of the mobile phase was achieved (the total time of HPLC-UV analysis was 5 min), which further allowed the cost of analysis to be reduced. In conclusion, the HPLC-UV method was convenient, precise and reproducible.

The developed method gives the possibility of detection of the analytes at low levels of concentrations. The proposed method was successfully applied to the determination of BUD and SULF in water and wastewater samples. As a result of the research, BUD and SULF were not found in the analyzed samples.

Author Contributions: Conceptualization, B.S. and M.H.; methodology, B.S. and M.H.; software, M.H.; validation, M.H. and A.G.; formal analysis, A.G. and M.H.; investigation, A.G. and M.H.; resources, B.S.; data curation, M.H.; writing—original draft preparation, M.H.; writing—review and editing, B.S.; visualization, M.H. and A.G.; supervision, M.H.; project administration, B.S.; funding acquisition, B.S.

Conflicts of Interest: The authors declare no conflict of interest

References

1. Semreen, M.H.; Shanableh, A.; Semerjian, L.; Alniss, H.; Mousa, M.; Bai, X.; Acharya, K. Simultaneous Determination of Pharmaceuticals by Solid-phase Extraction and Liquid Chromatography-Tandem Mass Spectrometry: A Case Study from Sharjah Sewage Treatment Plant. *Molecules* **2019**, *24*, 633. [CrossRef] [PubMed]
2. Wang, Y.; Tang, Y.; Moellmann, H.; Hochhaus, G. Simultaneous quantification of budesonide and its two metabolites, 6beta-hydroxybudesonide and 16alpha-hydroxyprednisolone, in human plasma by liquid chromatography negative electrospray ionization tandem mass spectrometry. *Biomed. Chromatogr.* **2003**, *17*, 158–164. [CrossRef] [PubMed]
3. Gazzotti, T.; Barbarossa, A.; Zironi, E.; Roncada, P.; Pietra, M.; Pagliuca, G. An LC-MS/MS method for the determination of budesonide and 16 alpha-hydroxyprednisolone in dog plasma. *Methodsx* **2016**, *3*, 139–143. [CrossRef] [PubMed]
4. Graham, G.G.; Pile, K.D. *Sulfasalazine and Related Drugs*; Encyclopedia of Inflammatory Diseases; Springer: New York, NY, USA, 2014; pp. 1–5.
5. Gupta, M.; Bhargava, H.N. Development and validation of a high-performance liquid chromatographic method for the analysis of budesonide. *J. Pharm. Biomed. Anal.* **2006**, *40*, 423–428. [CrossRef] [PubMed]
6. Naikwade, S.R.; Bajaj, A.N. Development of a validated specific HPLC method for budesonide and characterization of its alkali degradation product. *Can. J. Anal. Sci. Spectrosc.* **2008**, *53*, 113–122.
7. Deventer, K.; Mikulcikova, P.; Van Hoecke, H.; Van Eenoo, P.; Del-beke, F.T. Detection of budesonide in human urine after inhalation by liquid chromatography–mass spectrometry. *J. Pharm. Biomed. Anal.* **2006**, *42*, 474–479. [CrossRef]
8. Nilsson, K.; Andersson, M.; Beck, O. Phospholipid removal combined with a semi-automated 96-well SPE application for determination of budesonide in human plasma with LC-MS/MS. *J. Chrom. B-Anal. Technol. Biomed. Life Sci.* **2014**, *970*, 31–35. [CrossRef]
9. Szeitz, A.; Manji, J.; Riggs, K.W.; Thamboo, A.; Javer, A.R. Validated assay for the simultaneous determination of cortisol and budesonide in human plasma using ultra high performance liquid chromatography-tandem mass spectrometry. *J. Pharm. Biomed. Anal.* **2014**, *90*, 198–206. [CrossRef]
10. Joseph, S.; Menon, S.; Khera, S. Simultaneous determination of methotrexate and sulfasalazine in plasma by HPLC-DAD. *LC GC N. Am.* **2015**, *33*, 122–138.
11. Saini, B.; Bansal, G. Degradation study on sulfasalazine and a validated HPLC-UV method for its stability testing. *Sci. Pharm.* **2014**, *82*, 295–306. [CrossRef]
12. Patil, A.; Raheja, V.; Damre, A. Simultaneous analysis of intestinal permeability markers, caffeine, paracetamol and sulfasalazine by reverse phase liquid chromatography: A tool for standardization of rat everted gut sac model. *Asian J. Pharm. Clin. Res.* **2010**, *3*, 204–207.
13. Kwiecien, A.; Piatek, K.; Zmudzki, P.; Krzek, J. TLC-densitometric determination of sulfasalazine and its possible impurities in pharmaceutical preparations. *Acta Chrom.* **2015**, *27*, 623–635. [CrossRef]

14. Su, F.; Sun, Z.Q.; Liang, X.R. Development and validation of a quantitative NMR method for the determination of the commercial tablet formulation of sulfasalazine. *Curr. Pharm. Anal.* **2019**, *15*, 39–44. [CrossRef]
15. Ramezani, Z.; Dibaee, N. Determination of sulfasalazine in sulfasalazine tablets using silver nanoparticles. *Iran. J. Pharm. Sci.* **2012**, *8*, 129–134.
16. Gu, G.Z.; Xia, H.M.; Pang, Z.Q.; Liu, Z.Y.; Jiang, X.G.; Chen, J. Determination of sulphasalazine and its main metabolite sulphapyridine and 5-aminosalicylic acid in human plasma by liquid chromatography/tandem mass spectrometry and its application to a pharmacokinetic study. *J. Chrom. B* **2011**, *879*, 449–456. [CrossRef] [PubMed]
17. Grabic, R.; Fick, J.; Lindberg, R.H.; Fedorova, G.; Tysklind, M. Multi-residue method for trace level determination of pharmaceuticals in environmental samples using liquid chromatography coupled to triple quadrupole mass spectrometry. *Talanta* **2012**, *100*, 183–195. [CrossRef]
18. Gineys, N.; Giroud, B.; Vulliet, E. Analytical method for the determination of trace levels of steroid hormones and corticosteroids in soil, based on PLE/SPE/LC-MS/MS. *Anal. Bioanal. Chem.* **2010**, *397*, 2295–2302. [CrossRef]
19. Fiori, J.; Andrisano, V. LC-MS method for the simultaneous determination of six glucocorticoids in pharmaceutical formulations and counterfeit cosmetic products. *J. Pharm. Biomed. Anal.* **2014**, *91*, 185–192. [CrossRef]
20. Kasprzyk-Hordern, B.; Dinsdale, R.M.; Guwy, A.J. Multiresidue methods for the analysis of pharmaceuticals, personal care products and illicit drugs in surface water and wastewater by solid-phase extraction and ultra performance liquid chromatography-electrospray tandem mass spectrometry. *Anal. Bioanal. Chem.* **2008**, *391*, 1293–1308. [CrossRef]
21. Kasprzyk-Hordern, B.; Dinsdale, R.M.; Guwy, A.J. The effect of signal suppression and mobile phase composition on the simultaneous analysis of multiple classes of acidic/neutral pharmaceuticals and personal care products in surface water by solid-phase extraction and ultra performance liquid chromatography-negative electrospray tandem mass spectrometry. *Talanta* **2008**, *74*, 1299–1312.
22. Leong, M.I.; Huang, S.D. Dispersive liquid-liquid microextraction method based on solidification of floating organic drop combined with gas chromatography with electron-capture or mass spectrometry detection. *J. Chrom. A* **2008**, *1211*, 8–12. [CrossRef]
23. Ahmadi-Jouibari, T.; Fattahi, N.; Shamsipur, M. Rapid extraction and determination of amphetamines in human urine samples using dispersive liquid-liquid microextraction and solidification of floating organic drop followed by high performance liquid chromatography. *J. Pharm. Biomed. Anal.* **2014**, *94*, 145–151. [CrossRef]
24. Rahimi, A.; Hashemi, P. Development of a dispersive liquid-liquid microextraction method based on solidification of a floating organic drop for the determination of beta-carotene in human serum. *J. Anal. Chem.* **2014**, *69*, 352–356. [CrossRef]
25. Jian, Y.H.; Hu, Y.; Wang, T.; Liu, J.L.; Zhang, C.H.; Li, Y. Dispersive liquid-liquid microextraction based on solidification of floating organic drop with high performance liquid chromatography for determination of deca brominated diphenyl ether in surficial sediments. *Chin. J. Anal. Chem.* **2010**, *38*, 62–66. [CrossRef]
26. Hou, F.; Deng, X.; Jiang, X.; Yu, J. Determination of parabens in beverage samples by dispersive liquid-liquid microextraction based on solidification of floating organic droplet. *J. Chrom. Sci.* **2014**, *52*, 1332–1338. [CrossRef]
27. Yamini, Y.; Rezaee, M.; Khanchi, A.; Faraji, M. Dispersive liquid-liquid microextraction based on the solidification of floating organic drop followed by inductively coupled plasma-optical emission spectrometry as a fast technique for the simultaneous determination of heavy metals. *J. Chrom. A* **2010**, *1217*, 2358–2364. [CrossRef]
28. Shamsipur, M.; Fattahi, N.; Assadi, Y.; Sadeghi, M.; Sharafi, K. Speciation of As(III) and As(V) in water samples by graphite furnace atomic absorption spectrometry after solid phase extraction combined with dispersive liquid-liquid microextraction based on the solidification of floating organic drop. *Talanta* **2014**, *130*, 26–32. [CrossRef]
29. Moghadam, M.R.; Shabani, A.M.H.; Dadfarnia, S. Spectrophotometric determination of iron species using a combination of artificial neural networks and dispersive liquid-liquid microextraction based on solidification of floating organic drop. *J. Hazard. Mater.* **2011**, *197*, 176–182. [CrossRef]

Water **2019**, *11*, 1581

30. Li, Y.; Peng, G.; He, Q.; Zhu, H.; Al-Hamadani, S.M.Z.F. Dispersive liquid-liquid microextraction based on the solidification of floating organic drop followed by ICP-MS for the simultaneous determination of heavy metals in wastewaters. *Spectrochim. Acta Part A-Mol. Biomol. Spectrosc.* **2015**, *140*, 156–161. [CrossRef]

31. Al-Saidi, H.M.; Emara Adel, A.A. The recent developments in dispersive liquid–liquid microextraction for preconcentration and determination of inorganic analytes. *J. Saudi Chem. Soc.* **2014**, *18*, 745–761. [CrossRef]

32. Wang, H.M.; Jiang, X.H.; Lin, S.; Yi, H. Studies on determination of sulfasalazine and sulfapyridine in human plasma by HPLC and pharmacokinetics in human volunteers. *Chin. J. Antibiot.* **2013**, *38*, 223–226.

water

MDPI

Article

Rapid and Sensitive Analysis of Hormones and Other Emerging Contaminants in Groundwater Using Ultrasound-Assisted Emulsification Microextraction with Solidification of Floating Organic Droplet Followed by GC-MS Detection

Urszula Kotowska [1,*], Justyna Kapelewska [1], Adam Kotowski [2] and Ewelina Pietuszewska [1]

[1] Institute of Chemistry, University of Bialystok, Ciołkowskiego 1K, 15-245 Bialystok, Poland
[2] Faculty of Mechanical Engineering, Bialystok University of Technology, Wiejska 45C, 15-351 Białystok, Poland
* Correspondence: ukrajew@uwb.edu.pl; Tel.: +48-85-738-81-11; Fax: +48-85-747-01-13

Received: 24 June 2019; Accepted: 6 August 2019; Published: 8 August 2019

check for updates

Abstract: Ultrasound-assisted emulsification microextraction with solidification of floating organic droplet (USAEME-SFOD) has been applied to isolate hormones and other emerging contaminants from groundwater samples. Simultaneously with the extraction process, derivatization in the matrix was carried out using acetic anhydride. Quantification of studied organic pollutants was done through gas chromatography mass spectrometry (GC-MS). Hormones included β-estradiol (E2), estrone (E1), and diethylstilbestrol (DES). Other compounds belonged to groups of pharmaceuticals (diclofenac (DIC)), antiseptics (triclosan (TRC)), preservatives (propylparaben (PP) and butylparaben (BP)), sunscreen agents (benzophenone (BPH), and 3-(4-methylbenzylidene)camphor (3MBC)), repellents (N,N-diethyltoluamide (DEET)), industrial chemicals (bisphenol A (BPA), 4-t-octylphenol (4OP), 4-n-nonylphenol (4NP)). A non-toxic and inexpensive 1-undecanol was successfully used as the extraction solvent. Volume of extractant and derivatization agent, ionic strength, and time of extraction were optimized. Very low limits of detection (LoD) ranging from 0.01 to 5.9 ng/L were obtained. Recoveries ranged from 90% to 123%, with relative standard deviation being lower than 17%. The developed procedure was used to determine target compounds in groundwater collected at municipal waste landfills as well as in groundwater from wells distant from sources of pollution.

Keywords: ultrasound-assisted emulsification microextraction; solidification of floating organic droplet; gas chromatography-mass spectrometry; hormones; emerging contaminants; groundwater

---

1. Introduction

According to the most popular definition given by the United States Geological Survey, an emerging contaminant (EC) is "any synthetic or naturally occurring chemical that is not commonly monitored in the environment but has the potential to enter the environment and cause known or suspected adverse ecological and/or human health effects" [1,2].

NORMAN Network Europe (network of reference laboratories, research centers, and related organizations for the monitoring and biomonitoring of emerging environmental substances) differentiates between emerging substances and emerging pollutants. Emerging substances are defined as "substances that have been detected in the environment, but which are currently not included in routine monitoring programs at EU level and whose fate, behavior, and (eco)toxicological effects are not well understood". Emerging pollutants are defined as "pollutants that are currently not included in routine monitoring programs at the European level and which may be candidates for future regulation, depending on research on their (eco)toxicity, potential health effects and public perception and on monitoring data regarding their occurrence in the various environmental compartments" [3].

The EC group contains substances which vary in regard to their chemical structure, toxicity, and environmental behavior and includes, among others, industrial additives and by-products, personal care products, and human and animal pharmaceuticals (PPCPs), surfactants, flame retardants, hormones, and sterols [3]. This list is not complete and each year is extended with newly-detected artificial contaminants as well as naturally occurring trace compounds. In recent years it has expanded through the inclusion of nanomaterials and microplastic particles [4]. More than 2100 scientific studies published between 2007 and 2016 have proven that ECs, as biologically active compounds, exhibit a potential risk to humans, plants, and/or animals [2]. Many of them are able to alter the normal hormone function of wildlife and humans by mimicking or magnifying the effects of endogenous hormones, disrupting their synthesis and activity or the operation of hormone receptors. Strongest endocrine activity is demonstrated by natural and synthetic hormones which are introduced into the environment with insufficiently treated wastewater [5,6]. Estrone (E1) and β-estradiol (E2) are two common forms of natural estrogen secreted by the human body which are frequently found in aqueous environments. This is especially true in respect to E2 due to its widespread use as a contraceptive and in hormone replacement therapy (HRT). Another non-steroidal synthetic estrogen, diethylstilbestrol (DES), has been formerly used medicinally to prevent stillbirths and as a growth stimulant in feed given to poultry, cattle, and sheep. Despite the fact that its effects have been proven harmful it is still used in the treatment of breast and prostate cancer and, in some countries, in HRT [7,8]. Additionally, it has been shown that compounds making up some personal care products and industrial chemicals exhibit hormonal activity. The largest number of studies confirming their effect on the endocrine system is related to bisphenol A (BPA). It is suspected that estrogenic activity of BPA increases the risk of developing breast cancer in humans and may act as an antiandrogen causing feminizing side effects in men [9–12]. Other ECs, including, for example, propylparaben (PP), butylparaben (BP), 3-(4-methylbenzylidene)camphor (3MBC), 4-t-octylphenol (4OP), 4-n-nonylphenol (4NP), and triclosan (TRC), have also been confirmed or suspected of having endocrine disrupting effects on the functioning of estrogen, androgen, prolactin, insulin, or thyroid hormones. It is supposed that continuous exposure to some ECs causes increased birth weight in children, adult fat gain, diabetes, and may potentially affect eating disorders [9,10,13–15].

In recent years, micro-extraction in the liquid–liquid system (LLME) has become one of the most widely used techniques for the preparation of samples for EC determination [16,17]. The most commonly used LLME modification is a dispersive liquid–liquid microextraction (DLLME), developed in 2006 by Rezaee et al. [18]. This technique uses a ternary system consisting of an examined aqueous solution, an extraction solvent, and a dispersing solvent. The formation of the emulsion results in an unlimited contact area between two aqueous and organic phases producing prompt mass exchange. It is possible; however, to avoid the addition of a dispersing solvent to form the emulsion through the use of ultrasonic radiation. In 2008, Regueiro et al. [19] used ultrasound for the first time to support microextraction in a liquid–liquid system, developing the ultrasound-assisted emulsification-microextraction technique (USAEME). In liquid–liquid microextraction techniques it is usual to use solvents which are heavier than water, although these substances are mostly chlorinated. This is due to the fact that after the extraction process the microdroplet of the organic solvent in which the analyte is dissolved is located at the bottom of the tube. Its collection (e.g., with a syringe) is relatively easy, especially if test tubes with a conical bottom are used. In case of solvents that are lighter than water, it is much more difficult to separate the microdroplet of the organic solvent from the aqueous phase. One solution which facilitates work with such solvents uses the process of solidifying the floating solvent drop. The technique using the process of solidification of the floating organic drop microextraction (SFODME) was introduced in 2007 by Khalili Zanjani et al. [20]. In this technique, after the extraction process, the sample is placed in an ice bath to solidify the drop of the organic solvent in the upper part of the vessel in which the extraction is carried out. The drop is then transferred to a vial where it melts at room temperature. The solidification of the floating organic drop (SFOD) technique is combined with various liquid–liquid microextraction variants and used to determine both organic compounds and metal ions in various types of matrices [21–26].

In this study, a simple and sensitive analytical procedure for simultaneous determination of hormones and other EC compounds most frequently detected in surface water bodies is optimized. USAEME-SFOD is used for the separation and preconcentration of analytes, whereas GC-MS in the selected ion monitoring (SIM) mode is applied for their quantification. The influence of extraction and derivatization parameters (i.e., the type of organic solvent and solvent volume, extraction time, derivatization reagent volume, and amount of buffering salt) on analyte recovery is investigated. The developed USAEME-SFOD/GC-MS procedure was used to assay target compounds in groundwater samples from northeastern Poland. To our knowledge, this is the first study utilizing USAEME with a non-chlorinated solvent for the determination of studied compounds. The presence of compounds from the EC group in groundwater has been largely unexplored, especially when compared to surface and marine waters, and this work may provide important knowledge within this area.

2. Experimental

2.1. Reagents and Solvents

Materials, PP, BP, BPH, 3MBC, N,N-diethyltoluamide (DEET), OP, NP, TRC, BPA, diclofenac (DIC), E1, E2, DES, tricosane, and 1-undecanol, were obtained from Sigma-Aldrich (Darmstadt, Germany). Methanol and anhydrous disodium hydrogen phosphate (V) were provided by POCH (Gliwice, Poland). Acetic anhydride was purchased from Chempur (Piekary Śląskie, Poland). Stock solutions of each analyte (at 1 mg/mL of each) were prepared separately in methanol and stored at −18 °C for a period not exceeding one month. Working solutions were prepared by diluting the stock standard solution in methanol and storing them at −18 °C not longer than two weeks. Deionized water was obtained using a purification system (Milli-Q RG, Millipore, Burlington, MA, USA) and stored in glass bottles.

2.2. Groundwater Samples

Samples of groundwater were collected from six deep wells (the surface of water was at a depth of 15–46 m) and five shallow wells (the surface of the water was at a depth of 3–8 m) used for individual water supply. The wells are located in a region which is not directly affected by industrial sources of pollution. Samples of groundwater from contaminated sites (twelve) were collected from monitoring wells located in two municipal solid waste (MSW) landfill sites of non-hazardous and inert waste. All the sampling points were located in northeastern Poland. The geological structure of areas from which the samples were taken consists of clay–sand–gravel deposits. Samples were collected in glass bottles with Teflon-lined caps that were rinsed with the sample water on site and immediately carried to the laboratory where, upon arrival, they were filtered through a 0.45 μm pore size membrane filter and stored at −18 °C.

2.3. The Procedure of Ultrasound-Assisted Emulsification Microextraction with Solidification of Floating Organic Droplet (USAEME-SFOD) Coupled with In Situ Derivatization

For the simultaneous USAEME-SFOD and derivatization aliquots of 5 mL water samples were placed in 10 mL glass centrifuge tubes containing previously weighted 0.25 g of sodium hydrogen phosphate. The extraction solvent (1-undecanol, 20 μL) consisting of tricosane (5 μg/L) as an internal standard and the derivatization reagent (acetic anhydride, 250 μL) were added to the water sample and mixed. Immediately after, the tube was immersed in the ultrasonic Unitra Unima (Warsaw, Poland) water bath. Extractions were performed at 42 kHz ultrasound frequency and 230 W power for the duration of 8 min at room temperature. Emulsions were separated using centrifugation at

4000 rpm/min for 4 min in an MPW-250 Med. Instruments (Warsaw, Poland) laboratory centrifuge. After this process, the droplet of organic phase floated at the top of the test tube. The test tube was then cooled in an ice bath. After five minutes, 1-undecanol solidified and was transferred into a 150 μL micro vial with an integrated insert. It melted quickly at room temperature and GC-MS analysis of 1 μL of obtained solution was then performed as described in Section 2.4. The course of the USAEME-SFOD is shown in Figure 1.

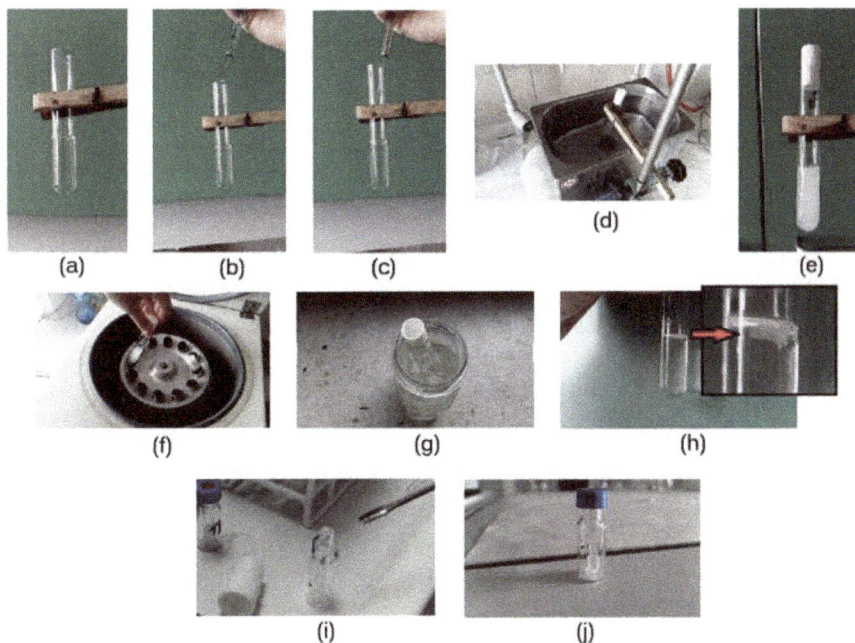

Figure 1. Stages of ultrasound-assisted emulsification-microextraction with solidification of the floating organic drop (USAEME-SFOD): (**a**) Placing test solution in a test tube; (**b**) adding acetic anhydride; (**c**) adding 1-undecanol; (**d**) subjecting the sample to sonication; (**e**) sample after sonication; (**f**) placing the tube in a centrifuge; (**g**) cooling in an ice bath; (**h**) a solid drop of solvent; (**i**) transferring the drop into a chromatography vial; (**j**) sample ready for GC-MS analysis.

2.4. GC-MS Conditions

Analysis was performed with an HP 6890 gas chromatograph coupled with a MSD5973 mass spectrometric detector and an HP 7673 autosampler (Agilent Technologies, Santa Clara, CA, USA). This device was equipped with an HP-5MS (5% phenylmethylsiloxane) column (size 30 m length × 0.25 mm; i.e., coated with 0.25 μm film thickness) and split/splitless injector. The injector was set to work in the splitless mode. Helium of 99.999% purity was used as a carrier gas at a flow rate of 1 mL/min. The injector temperature was set at 250 °C. The oven temperature program started from 160 °C and increased at increments of 2 °C/min to 170 °C (held for 2 min), 6.44 °C/min to 226 °C (held for 1 min), 10 °C/min to 233 °C (held for 2.5 min) and 10 °C/min to 300 °C (held for 4 min). The total run time was 26 min with the solvent delay time reaching 5.7 min. The MS detector worked in selected ion monitoring (SIM) mode. The electron impact source temperature was 230 °C with electron energy of 70 eV. The quadrupole temperature was 150 °C and the GC interface temperature was 280 °C. The chromatogram obtained during the GC-MS analysis of BPH, DEET, 3MBC, BPA, PP, BP, DES, DIC, OP, NP, E1, E2, and TRC together with the internal standard after USAEME-SFOD with in-situ acetylation is presented in Figure 2. The MS spectra of target compounds are given in Figure S1

(Supplementary Material). The chromatographic parameters, molecular weights of target compounds, together with quantification and identification ions are shown in Table 1. Based on the registered chromatograms, the relative areas of the chromatographic peaks were determined by dividing the values of the obtained peak areas of tested compounds by the peak area of the tricosane.

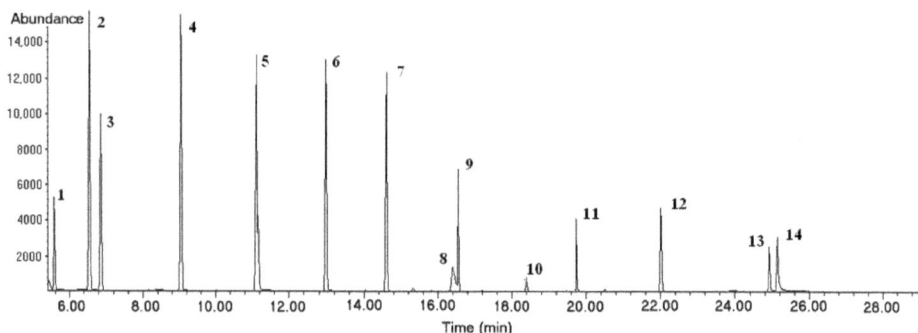

Figure 2. Chromatogram obtained during the GC-MS analysis: (1) N,N-diethyltoluamide (DEET), (2) benzophenone (BPH), (3) propylparaben (PP), (4) butylparaben (BP), (5) 4-t-octylphenol (4OP),(6) 4-n-nonylphenol (4NP), (7) 3-(4-methylbenzylidene)camphor (3MBC), (8) diclofenac (DIC), (9) triclosan (TRC) (10) tricosane (internal standard), (11) bisphenol A (BPA), (12) diethylstilbestrol (DES), (13) estrone (E1), (14) β-estradiol (E2).

Table 1. Retention times, molecular weights of target compounds, together with quantification and identification ions.

Analyte	Group	Formula	Molar Weight (MW), (g/mol)	Chemical Abstracts Service Number	Retention Time (min)	Quantification and Identification Ion (m/z)
E2	Natural steroid hormone	$C_{18}H_{24}O_2$	272	50-28-2	25.05	43, 146, 272
E1	Natural steroid hormone	$C_{18}H_{22}O_2$	270	53-16-7	24.81	185, 270, 272
DES	Artificial non-steroid hormone	$C_{18}H_{20}O_2$	268	56-53-1	21.87	268, 310, 352
DIC	Non-steroidal anti-inflammatory drug	$C_{14}H_{11}Cl_2NO_2$	296	15,307-79-6	16.24	214, 242, 277
PP	Preservative	$C_{10}H_{12}O_3$	180	94-13-3	6.71	121, 138, 180
BP	Preservative	$C_{11}H_{14}O_3$	194	94-26-8	8.90	121, 138, 194
BPH	UV filter	$C_{13}H_{10}O$	182	119-61-9	6.41	77, 105, 182
3MBC	UV filter	$C_{18}H_{22}O$	254	36,861-47-9	14.47	128, 171, 254
DEET	Repellent	$C_{12}H_{17}NO$	191	134-62-3	5.46	91, 119, 190
4OP	Nonionic surfactant	$C_{14}H_{22}O$	206	1806-26-4	10.96	43, 107, 206
4NP	Nonionic surfactant	$C_{15}H_{24}O$	220	84,852-15-3	12.85	43, 107, 220
BPA	Substrate in the production of plastics	$C_{15}H_{16}O_2$	228	80-05-7	19.53	213, 228, 270
TRC	Antiseptic	$C_{12}H_7Cl_3O_2$	289.5	3380-34-5	16.39	218, 288, 290

3. Results and Discussion

3.1. Optimization of Extraction and Derivatization Procedure

3.1.1. Selection of an Extraction Solvent

The efficiency of the liquid–liquid extraction process depends on the physico-chemical properties, such as water solubility (it has to be as low as possible) and its polarity or affinity to the isolated compounds (it has to be as high as possible), of the solvent used. Additionally, the selected organic solvent should exhibit low volatility and toxicity, and proper chromatographic behavior. The additional feature of the chosen solvent required for its application in USAEME is the ability to form an emulsion during the extraction procedure. To make solidification of the solvent droplet possible the melting point of the solvent should be near room temperature. Four solvents were tested (Table 2) for their usability in the isolation of studied compounds. Based on an analysis of their physicochemical properties and on preliminary tests, 1-dodecanol and n-hexadecane were eliminated. The former is characterized by a melting temperature that is too high and becomes solid during extraction, which would require conducting the experiments at an elevated temperature. The high boiling point of n-hexadecane masked a large part of the obtained GC-MS chromatogram making determination of some analytes impossible. As the extraction efficiency of 1-undecanol and 2-dodecanol were similar, 1-undecanol was chosen for further studies due to its lower boiling point allowing the GC-MS determination of more endocrine disrupting compounds (less chromatogram coverage).

Table 2. Solvents tested for target compounds extraction by USAEME-SFOD.

Solvent	CAS	Molar Mass (g/mol)	Melting Point (°C)	Boiling Point (°C)	Density (g/mL)
1-Undecanol	112-42-5	172.31	13	243	0.830
2-Dodecanol	10,203-28-8	186.34	18	250	0.829
1-Dodecanol	112-53-8	186.34	24	259	0.831
n-Hexadecane	544-76-3	226.41	18	286.8	0.773

3.1.2. Selection of Solvent Volume

The analyte extraction efficiency strongly depends on the volume of the solvent used. Usually, a decrease in the volume of the organic phase produces higher enrichment of the analyte. The use of as small as possible amount of organic phase improves the limits of detection (LoD) and quantification (LoQ) of the applied detection technique. What is more, using the smallest possible amounts of solvents conforms to the guidelines of "green chemistry".

In order to obtain the highest extraction efficiency of the USAEME-SFOD procedure, the volume of extraction solvent was optimized. For this purpose, different volumes of 1-undecanol within the range of 20–70 µL were examined (Figure 3). The obtained results showed that a decrease in the solvent volume resulted in the elevation of peak areas of analyzed compounds. It was determined that the use of 20 µL of 1-undecanol, the smallest volume making the introduction of the sample into the chromatograph with an autosampler possible, allowed the collection of 10 to 12 µL of solvent after the extraction process. Therefore, a volume 20 µL of 1-undecanol was deemed to be optimal in further studies.

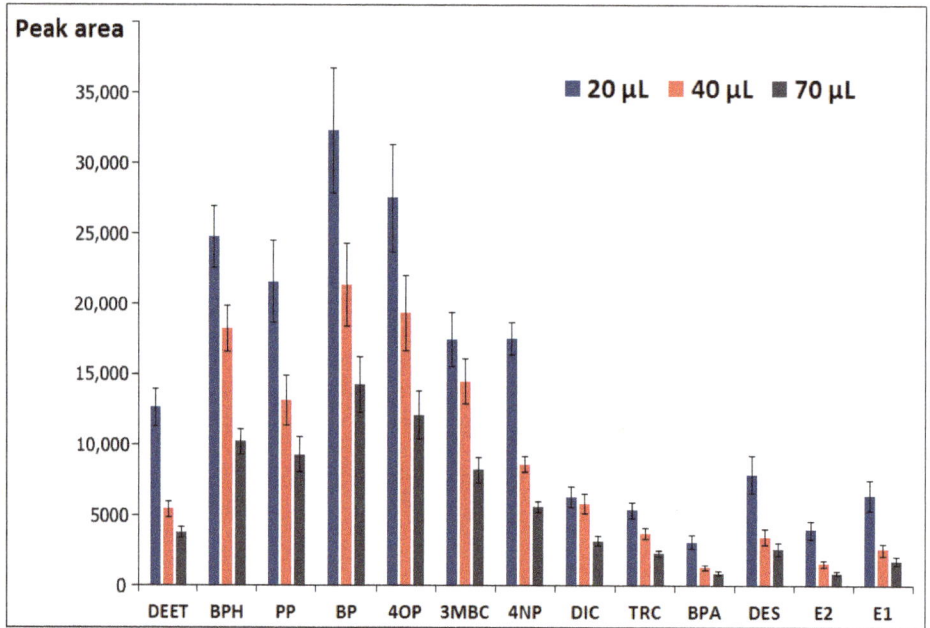

Figure 3. The influence of the 1-undecanol volume on isolation efficiency.

3.1.3. Effect of Derivatization Reagent Volume

Ten of the thirteen examined endocrine compounds contain a hydroxyl or phenol group in their molecule, requiring appropriate conversion increasing volatility and thermal stability to improve their adaptation to GC-MS analysis. For this purpose, derivatization in the matrix using acetic anhydride was chosen and was carried out simultaneously with the extraction procedure. Besides providing chromatographic advantages, acetylation also improves the sensitivity of the method, as the extraction efficiency of the acetates formed in the reaction is much higher than the yield of phenol extraction [27]. In-situ derivatization with acetic anhydride is simple as well as fast, having little effect on the duration and complexity of the analytical procedure [28]. The acetylation reaction has been confirmed through the appearance within the spectrum a peak of 42 (or 84) units higher than the molecular weight of the compound being determined (see Figure S1, Supplementary Material). The effect of the volume of acetic anhydride on the relative peak area was studied in the range 60–250 µL (Figure 4). The results indicated that the volume of acetic anhydride equal to 250 µL is optimum, providing the highest efficiency of extraction. The use of large volumes of acetic anhydride is advantageous because it provides effective derivatization in the case of environmental samples with an unknown concentration of analytes. In addition, during the isolation, a reaction between the acetic anhydride and the solvent, which is the primary alcohol, undoubtedly takes place. A large stoichiometric excess of acetic anhydride means that the occurring reaction does not affect the process of derivatization of analytes, which is also confirmed by studies described in the literature [28].

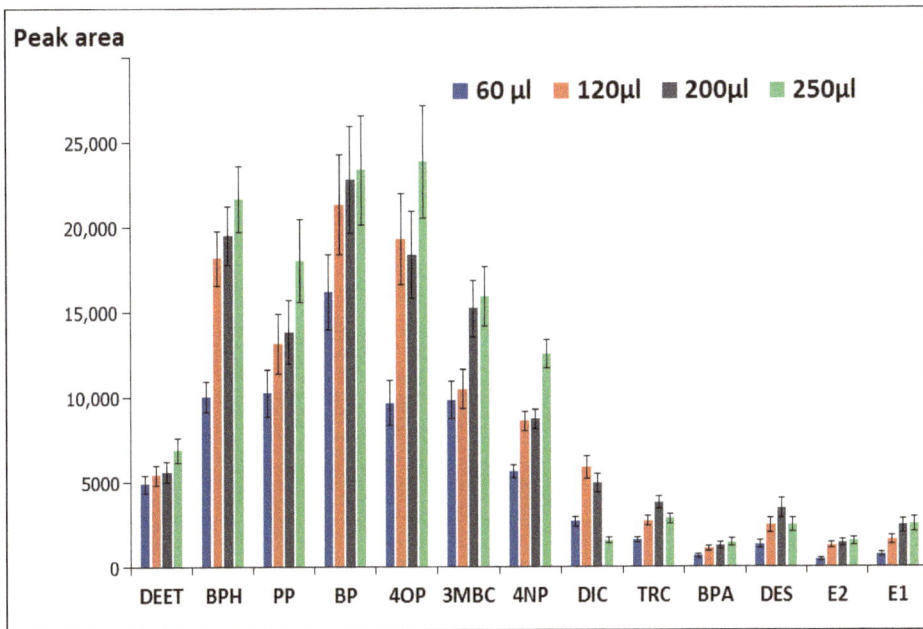

Figure 4. The influence of the volume of acetic anhydride on isolation efficiency.

3.1.4. Effect of Type and Amount of Buffering Salt

The acylation reaction requires the presence of buffer salt with sodium hydrogen carbonate, being most frequently used for this purpose. However, this substance turned out to be unsuitable for the USAEME-SFOD procedure, since the use of $NaHCO_3$ resulted in the appearance of carbon dioxide bubbles which interfered with the agglomeration of the organic phase and was replaced with sodium hydrogen phosphate. The addition of salt to the solution also causes a salting-out effect which positively affects the efficiency of the extraction. To find the optimal concentration of salt, a series of experiments was performed using solutions with salt concentrations between 0% and 10%. Figure 5 shows the influence of sodium hydrogen phosphate concentration on the extraction efficiency of target compounds. It can be seen that the best results were obtained with the 10% solution and this concentration was used in subsequent experiments.

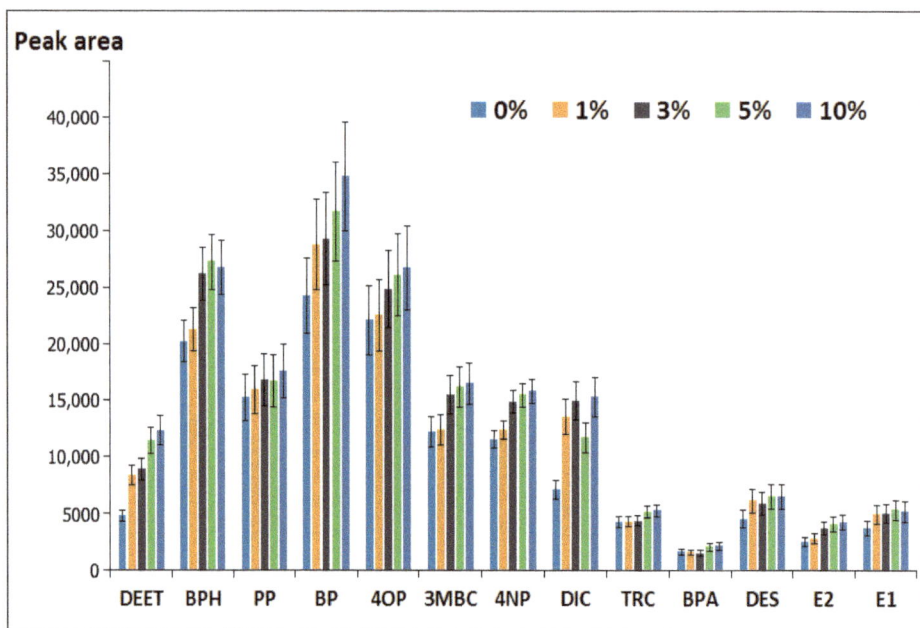

Figure 5. The influence of sodium hydrogen phosphate concentration on isolation efficiency.

3.1.5. Effect of the Simultaneous Derivatization/Extraction Time

Literary data indicates that acetylation is a fast process which can be conducted with 100% efficiency within a time of two minutes [29]. Therefore, the rate of simultaneous derivatization and extraction is mainly related to the time required to achieve equilibrium in the distribution of the analyte between the aqueous and organic phases, which is of a great significance in all extraction procedures. When it comes to USAEME, extraction time is considered as the time between injection of the extraction solvent and the end of the sonication stage [30]. In order to select a time interval that ensures the highest efficiency of extraction, the isolation of studied compounds was done using varying sonication times ranging from 2 to 15 min (Figure 6). For the majority of the examined compounds the time of 8 min was enough to establish equilibrium between the aqueous and organic phases, a fact that can be clearly seen on the graph. In case of BPH and 3MBC, equilibrium concentrations in the 1-undecanol–water system were achieved after only two minutes of extraction. For DIC, TRC, and E1, slightly higher extraction efficiency was registered after 12 min of sonication. Since it is known that the achievement of equilibrium is not necessary for repetitive extraction in the liquid–liquid system, a time of 8 min was chosen as sufficient to perform the sample preparation procedure. The choice of this extraction time was also related to the fact that, for some acetylated compounds, a decrease of concentration in the organic phase was observed with the extension of the sonication time. The negative effect of extraction time elongation on extraction efficiency suggests that the derivatives may have slowly hydrolyzed to their free form after coming in contact with the aqueous phase.

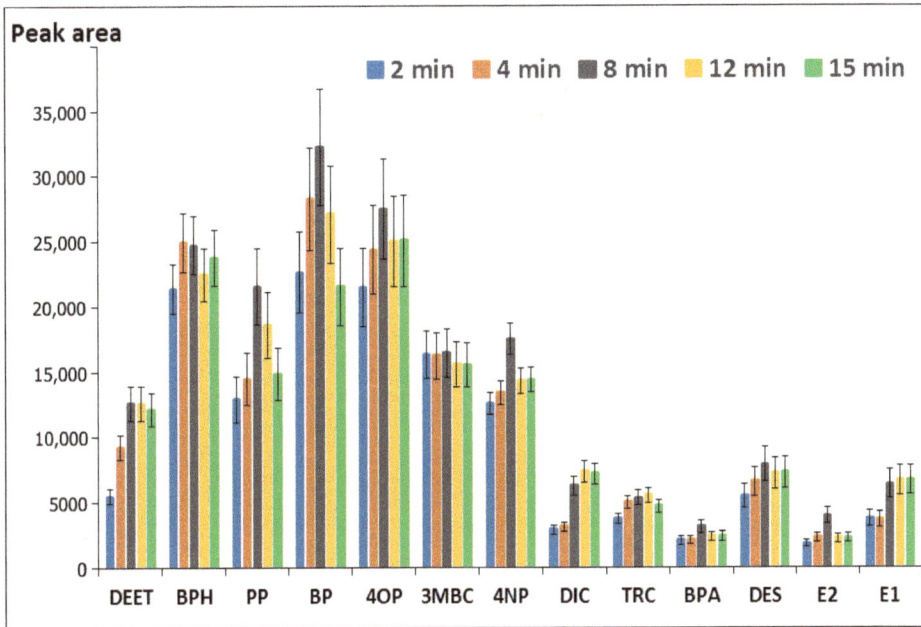

Figure 6. The influence of the extraction time on isolation efficiency.

3.2. Method Validation

Method validation was done using real groundwater samples, for which the absence of determined compounds was confirmed, as the sample matrix. Table 3 presents an overview of the method's linearity studies. Table 4 shows limits of detection and quantification repeatability and trueness of the developed method as well as a comparison with values obtained when USAEME is used with solvent having a density higher than water. Within the studied concentration range 0.001–10 μg/L, calibration graphs were linear and corresponded with the expected concentrations in groundwater. It has been proven that conducting determinations based on a standard curve with a wide dynamic range (above two orders of magnitude) leads to substantial errors at low concentrations. In order to improve the accuracy of the determinations, the dynamic range was divided into two (low and high) operating scopes—0.001–0.05 μg/L and 0.05–10 μg/L—and validation of both scopes was performed. As can be seen in Table 3, the equations of the calibration curves for the low and high operating scopes differ, confirming the correctness of the approach used. Good linearity with coefficient of determination (r^2) ranging from 0.990 to 0.999 for the high operating scope, and from 0.985 to 0.997 for the low operating scope, was obtained. Relative standard deviation (RSD) of the determination ranged from 6.7% to 16.8%, depending on the compound being analyzed. Analyte limits of quantification defined as a signal to noise ratio (S/N) equal to 10 ranged from 0.05 to 19.5 ng/L; the limits of detection, defined as S/N ratio equal to 3, were between 0.01 and 5.9 ng/L. Recovery obtained from groundwater samples spiked at concentration levels of 0.03 μg/L were between 93% and 123%, while for samples spiked at concentration 4 μg/L were between 90% and 112%. A comparison of parameters obtained through the utilization of the USAEME-SFOD/GC-MS method developed using 1-undecanol as an extractant, with those attained by applying the USAEME/GC-MS method using chloroform as an extractant [31], shows that the accuracy and precision of both methods are similar, while the sensitivity of the method developed in this work is much better, primarily resulting from the smaller volume of 1-undecanol (20 μL) used for extraction compared to chloroform (70 μL), lower water solubility of 1-undecanol (5.7 mg/L) than chloroform (0.8 g/L), and, most likely, a higher solubility of target compounds in 1-undecanol.

Table 3. Overview of the method's linearity studies.

Analyte	Equation of the Calibration Curve *		Coefficient of Determination (r²) *	
	Range I (0.001–0.05 µg/L)	Range II (0.05–10 µg/L)	Range I (0.001–0.05 µg/L)	Range II (0.05–10 µg/L)
E2	-	y = 1940.2x − 144.7	-	0.9982
E1	y = 2835.0x + 254.5	y = 5214.1x − 570.5	0.9970	0.9965
DES	y = 7036.5x + 234.7	y = 5105.1x − 861.9	0.9949	0.9974
DIC	y = 3102.3 + 292.2	y = 1490.7x + 500.0	0.9894	0.9929
PP	y = 6120.7x + 309.9	y = 5807.8x + 799.4	0.9891	0.9993
BP	y = 30,653.4x + 1269.4	y = 9395.7x + 2330.7	0.9872	0.9960
BPH	y = 191,925.7x + 7247.8	y = 7793.9x + 16284.3	0.9944	0.9993
3MBC	y = 4671.2x + 128.3	y = 2580.6x + 113.7	0.9901	0.9978
DEET	y = 8894.3x + 1231.8	y = 3955.3x + 1761.6	0.9853	0.9988
4OP	y = 31,697.4x + 658.1	y = 18,968.0x − 298.7	0.9935	0.9962
4NP	y = 10,575.9x + 236.2	y = 8932.6x + 293.0	0.9890	0.9904
BPA	y = 119,133.7x + 3285.2	y = 10,428.7x + 6435.3	0.9916	0.9974
TRC	y = 6070.4x + 215.5	y = 4535.3x + 674.8	0.9869	0.9994

* Real groundwater was used as the sample matrix in method validation.

Table 4. Limits of detection and quantification, repeatability, and trueness of the developed method, and its comparison with values obtained when the solvent with a density higher than water (chloroform) is used.

Analyte	USAEME-SFOD/GC-MS (This Work) *					USAEME/GC-MS (From Literature [31]) *			
	Recovery (%)		RSD (%)	LoD (ng/L)	LoQ (ng/L)	Recovery (%)	RSD (%)	LoD (ng/L)	LoQ (ng/L)
	0.03 µg/L	4.00 µg/L				1 µg/L			
E2	121.3	90.2	15.5	5.9	19.5	103	14.3	130.73	435.77
E1	-	111.5	16.7	1.54	5.10	103	10.3	8.92	29.73
DES	96.7	90.1	16.8	0.04	0.12	101	13.7	88.36	294.54
DIC	93.3	99.1	11.4	0.17	0.56	110	12.1	149.55	498.48
PP	116.7	95.5	13.5	0.05	0.16	97	12.4	23.62	47.24
BP	120.0	92.2	13.8	0.04	0.15	126	14.0	10.21	34.03
BPH	110.0	103.7	8.9	0.03	0.11	105	16.2	2.97	9.90
3MBC	120.7	93.7	11.0	0.04	0.15	116	11.2	3.01	10.00
DEET	113.8	99.5	10.4	0.02	0.05	117	17.3	1.50	4.99
4OP	113.3	93.1	13.9	0.02	0.07	105	10.3	1.50	5.00
4NP	112.9	97.7	6.7	0.01	0.05	101	14.7	2.94	9.80
BPA	123.1	94.7	15.8	0.04	0.12	107	15.6	1.49	4.98
TRC	123.3	96.1	10.2	0.04	0.14	134	15.8	2.48	8.26

RSD—relative standard deviation; LoD—limit of detection; LoQ—limit of quantification. * Real groundwater was used as the sample matrix in method validation.

3.3. Groundwater Analysis

In order to complete the validation of the proposed procedure, it was applied for the determination of PP, BP, BPH, 3MBC, DEET, OP, NP, TRC, BPA, DIC, E1, E2, and DES in groundwater samples from deep wells and shallow wells used for individual water supply, as well as groundwater from monitoring wells located in municipal solid waste landfill sites. Analysis results are shown in Table 5. One hormone (E1) and nine other ECs (with the exception of 3MBC) were identified in the groundwater samples in concentrations higher than LoD. The most frequently detected compounds were BPA (detection frequency 78%), DEET (detection frequency 61%), and BPH (detection frequency 57%). Literature information from other studies shows that BPA and DEET were identified as the most commonly detected ECs in groundwater [32–36] with BPH being previously identified as one of the most widespread ECs in groundwater located under MSW landfills [31,37–39].

Table 5. The ECs concentrations (ng/L; range and median) in groundwater samples from deep wells, shallow wells, and monitoring wells located in two municipal solid waste (MSW) landfill sites with detection frequencies.

Analyte	Groundwater Samples from Drilling Wells (NS = 6)			Groundwater Samples from Shallow Wells (NS = 5)			Groundwater Samples from MSW Monitoring Wells (NS = 12)		
	Range (ng/L)	Median (ng/L)	d.f.	Range (ng/L)	Median (ng/L)	d.f.	Range (ng/L)	Median (ng/L)	d.f.
E2	n.d.	-	0	n.d.	-	0	n.d.	-	0
E1	n.d.	-	0	n.d.	-	0	n.d.–309	107	2
DES	n.d.	-	0	n.d.	-	0	n.d.	-	0
DIC	n.d.	-	0	n.d.	-	0	n.d.–312	280	3
PP	n.d.	-	0	n.d.	-	0	n.d.–0.5	-	1
BP	n.d.	-	0	n.d.	-	0	n.d.–0.2	0.2	2
BPH	n.d.	-	0	n.d.–124	-	1	0.5–3300	33	12
3MBC	n.d.	-	0	n.d.	-	0	n.d.	-	0
DEET	n.d.–2	2	2	n.d.–21	6	2	n.d.–3	2	10
4OP	n.d.	-	0	n.d.–17	11	2	n.d.–25	10	3
4NP	n.d.	-	0	n.d.–8	-	1	n.d.–9	7	2
BPA	n.d.–98	53	2	n.d.–689	124	4	0.2–1050	79	12
TRC	n.d.	-	0	n.d.	-	0	n.d.–38	28	3

NS = number of samples, d.f. = detection frequencies, number of samples with concentration higher than LoD; n.d—not detected.

All ten compounds, in concentrations ranging from 0.2 to 3300 ng/L, were detected in groundwater samples from MSW monitoring wells, with BPH and BPA present in all samples. Insufficient insulation of fields on which waste is stored and lack of completely watertight installations used for collecting landfill leachate are, in this case, the main source of the above mentioned compounds. In shallow wells, five compounds (BPH, DEET, 4OP, 4NP, BPA), in concentrations from 0.1 to 689 ng/L, were detected. Water in shallow wells originate from the near-surface usable aquifer associated with the sandy fluvioglacial and glacial sediments of the Upper Pleistocene [40]. The presence of ECs in these waters results from it being fed directly by precipitation and meltwater. DEET and BPA, in concentrations ranging 0.1–98 ng/L, were both detected in two deep wells. Deep-well water comes from the first main usable deep water aquifer called the inter-renewable level. It occurs in fluvioglacial works of the oldest glaciation stages in Central Poland and in gravel and river sand interstadials [40]. The results indicate that the two deep wells in which ECs were detected are affected by seep water from shallow reservoirs coming in contact with surface waters.

4. Conclusions

- New analytical methodology based on ultrasound-assisted emulsification microextraction with solidification of organic drop followed by GC-MS determination has been proposed for the determination of three hormones and ten other ECs having a high environmental impact. Scrutiny of the available literary sources showed that the present work is the first to describe the combination of the USAEME and SFOD methods for the extraction of target compounds from any matrix.
- High sensitivity of the developed procedure and satisfactory precision and accuracy enabled its use for the determination of ECs in groundwater samples, which are usually characterized by low contamination by anthropogenic compounds.
- Analyses of groundwater have shown that even their deep seams can be contaminated with compounds derived from industrial and everyday human activity. This is a particularly worrying phenomenon since these resources are utilized as sources of high-quality drinking water.

Supplementary Materials: The following are available online at http://www.mdpi.com/2073-4441/11/8/1638/s1, Figure S1: The MS spectra of target compounds registered after USAEME-SFOD with in-situ acetylation.

Author Contributions: Conceptualization, U.K.; methodology, U.K. and J.K.; investigation, U.K., J.K. and E.P.; writing—original draft preparation, U.K. and A.K.; writing—review and editing, U.K. and A.K.; visualization, A.K.

Funding: This research was funded by the Institute of Chemistry, University of Bialystok.

Conflicts of Interest: The authors declare no conflicts of interest.

References

1. Website United States Geological Survey. Available online: http//:www.toxics.usgs.gov/regional/emc/ (accessed on 4 March 2019).
2. Alphenaar, P.A.; van Houten, M. *Insight in Emerging Contaminants in Europe*; Witteveen Bos: Deventer, The Netherlands, 2016.
3. Norman Network of Reference Laboratories. Research Centres and Related Organisations for Monitoring of Emerging Environmental Substances. Available online: https://www.norman-network.net (accessed on 4 March 2019).
4. Richardson, S.D.; Kimura, S.Y. Water analysis: Emerging contaminants and current issues. *Anal. Chem.* **2016**, *88*, 546–582. [CrossRef] [PubMed]
5. Barber, L.B.; Vajda, A.M.; Douville, C.H.; Norris, D.O.; Writer, J.H. Fish Endocrine disruption responses to a major wastewater treatment facility upgrade. *Environ. Sci. Technol.* **2012**, *46*, 2121–2131. [CrossRef] [PubMed]
6. Laurenson, J.P.; Bloom, R.A.; Page, S.; Sadrieh, N. Ethinyl estradiol and other human pharmaceutical estrogens in the aquatic environment: A review of recent risk assessment data. *AAPS J.* **2014**, *16*, 299–310. [CrossRef] [PubMed]
7. Zhu, L.; Xiao, L.; Xia, Y.; Zhou, K.; Wang, H.; Huang, M.; Ge, G.; Wu, Y.; Wu, G.; Yang, L. Diethylstilbestrol can effectively accelerate estradiol-17-O-glucuronidation, while potently inhibiting estradiol-3-O-glucuronidation. *Toxicol. Appl. Pharmacol.* **2015**, *283*, 109–116. [CrossRef] [PubMed]
8. Hao, C.J.; Cheng, X.J.; Xia, H.F.; Ma, X. The endocrine disruptor diethylstilbestrol induces adipocyte differentiation and promotes obesity in mice. *Toxicol. Appl. Pharmacol.* **2012**, *263*, 102–110. [CrossRef]
9. Routledge, J.E.; Parker, J.; Odum, J.; Ashby, J.; Sumpter, J.P. Some alkyl hydroxybenoate preservatives (parabens) are estrogenic. *Toxicol. Appl. Pharmacol.* **1998**, *153*, 12–19. [CrossRef] [PubMed]
10. Gogoi, A.; Mazumder, P.; Tyagi, V.K.; Tushara Chaminda, G.G.; Kyoungjin Ane, A.; Kumar, M. Occurrence and fate of emerging contaminants in water environment: A review. *Groundw. Sustain. Dev.* **2018**, *6*, 169–180. [CrossRef]
11. Watson, C.; Zoeller, T.; Belcher, S. In vitro molecular mechanisms of bisphenol A action. *Reprod. Toxicol.* **2007**, *24*, 178–198.
12. Fenichel, P.; Chevalier, N.; Brucker-Davis, F. Bisphenol A: An endocrine and metabolic disruptor. *Ann. Endocrinol.* **2013**, *74*, 211–220. [CrossRef]
13. Boas, M.; Feldt-Rasmussen, U.; Main, K.M. Thyroid effects of endocrine disrupting chemicals. *Mol. Cell. Endocrinol.* **2012**, *355*, 240–248. [CrossRef]
14. Veiga-Lopez, A.; Pu, Y.; Gingrich, J.; Padmanabhan, V. Obesogenic endocrine disrupting chemicals: Identifying knowledge gaps. *Trends Endocrinol. Metab.* **2018**, *29*, 607–625. [CrossRef] [PubMed]
15. Walley, S.N.; Roepke, T.A. Perinatal exposure to endocrine disrupting compounds and the control of feeding behavior—An overview. *Horm. Behav.* **2018**, *101*, 22–28. [CrossRef] [PubMed]
16. Hashemi, B.; Zohrabi, P.; Kim, K.H.; Shamsipur, M.; Deep, A.; Hong, J. Recent advances in liquid-phase microextraction techniques for the analysis of environmental pollutants. *Trends Anal. Chem.* **2017**, *97*, 83–95. [CrossRef]
17. Seidi, S.; Rezazadeh, M.; Yamini, Y. Pharmaceutical applications of liquid-phase microextraction. *Trends Anal. Chem.* **2018**, *108*, 296–305. [CrossRef]
18. Rezaee, M.; Assadi, Y.; Hosseini, M.R.M.; Aghaee, E.; Ahmadi, F.; Berijani, S. Determination of organic compounds in water using dispersive liquid–liquid microextraction. *J. Chromatogr. A* **2006**, *1116*, 1–9. [CrossRef] [PubMed]
19. Regueiro, J.; Llompart, M.; Garcia-Jares, C.; Garcia-Monteagudo, J.C.; Cela, R. Ultrasound-assisted emulsification–microextraction of emergent contaminants and pesticides in environmental waters. *J. Chromatogr. A* **2008**, *1190*, 27–38. [CrossRef]
20. Khalili Zanjani, M.R.; Yamini, Y.; Shariati, S.; Jönsson, J.A. A new liquid-phase microextraction method based on solidification of floating organic drop. *Anal. Chim. Acta.* **2007**, *585*, 286–293. [CrossRef]

21. Yang, D.; Yang, Y.; Li, Y.; Yin, S.; Chen, Y.; Wang, J.; Xiao, J.; Sun, C. Dispersive liquid–liquid microextraction based on solidification of floating organic drop combined with high performance liquid chromatography for analysis of 15 phthalates in water. *J. AOAC Int.* **2019**, *102*, 942–951. [CrossRef]

22. Jafariyan, R.; Shabani, A.M.H.; Dadfarnia, S.; Tafti, E.N.; Shirani, M. Dispersive liquid–liquid microextraction based on solidification of floating organic drop as an efficient preconcentration method for spectrophotometric determination of aluminium. *Anal. Bioanal. Chem. Res.* **2019**, *6*, 289–299.

23. Ezoddin, M.; Adlnasab, L.; Kaveh, A.A.; Karimi, M.A. Ultrasonically formation of supramolecular based ultrasound energy assisted solidification of floating organic drop microextraction for preconcentration of methadone in human plasma and saliva samples prior to gas chromatography–mass spectrometry. *Ultrason. Sonochem.* **2019**, *50*, 182–187. [CrossRef]

24. Guiñez, M.; Bazan, C.; Martinez, L.D.; Cerutti, S. Determination of nitrated and oxygenated polycyclic aromatic hydrocarbons in water samples by a liquid–liquid phase microextraction procedure based on the solidification of a floating organic drop followed by solvent assisted back-extraction and liquid chromatography–tandem mass spectrometry. *Microchem. J.* **2018**, *139*, 164–173.

25. Asadi, M. Determination of ochratoxin A in fruit juice by high-performance liquid chromatography after vortex-assisted emulsification microextraction based on solidification of floating organic drop. *Mycotoxin Res.* **2018**, *34*, 15–20. [CrossRef] [PubMed]

26. Aydin, I.; Chormey, D.S.; Budak, T.; Firat, M.; Turak, F.; Bakirdere, S. Development of an accurate and sensitive analytical method for the determination of cadmium at trace levels using dispersive liquid–liquid microextraction based on the solidification of floating organic drops combined with slotted quartz tube flame atomic absorption spectrometry. *J. AOAC Int.* **2018**, *101*, 843–847. [PubMed]

27. Coutts, R.T.; Hargesheimer, E.E.; Pasutto, F.M. Gas chromatographic analysis of trace phenols by direct acetylation in aqueous solution. *J. Chromatogr. A* **1979**, *179*, 291–299. [CrossRef]

28. Faraji, H.; Tehrani, M.S.; Husain, S.W. Pre-concentration of phenolic compounds in water samples by novel liquid–liquid microextraction and determination by gas chromatography–mass spectrometry. *J. Chromatogr. A* **2009**, *1216*, 8569–8574. [CrossRef]

29. Casado, J.; Nescatelli, R.; Rodríguez, I.; Ramil, M.; Marini, F.; Cela, R. Determination of benzotriazoles in water samples by concurrent derivatization–dispersive liquid–liquid microextraction followedby gas chromatography–mass spectrometry. *J. Chromatogr. A* **2014**, *1336*, 1–9. [CrossRef] [PubMed]

30. Ma, J.J.; Du, X.; Zhang, J.W.; Li, J.C.; Wang, L.Z. Ultrasound-assisted emulsification-microextraction combined with flame atomic absorption spectrometry for determination of trace cadmium in water samples. *Talanta* **2009**, *80*, 980–984. [CrossRef]

31. Kapelewska, J.; Kotowska, U.; Karpińska, J.; Kowalczuk, D.; Arciszewska, A.; Świrydo, A. Occurrence, removal, mass loading and environmental risk assessment of emerging organic contaminants in leachates, groundwaters and wastewaters. *Microchem. J.* **2018**, *137*, 292–301. [CrossRef]

32. Lapworth, D.J.; Baran, N.; Stuart, M.E.; Ward, R.S. Emerging organic contaminants in groundwater: A review of sources, fate and occurrence. *Environ. Pollut.* **2012**, *163*, 287–303. [CrossRef]

33. Bexfield, L.M.; Toccalino, P.L.; Belitz, K.; Foreman, W.T.; Furlong, E.T. Hormones and pharmaceuticals in groundwater used as a source of drinking water across the United States. *Environ. Sci. Technol.* **2019**, *53*, 2950–2960. [CrossRef]

34. Schulze, S.; Zahn, D.; Montes, R.; Rodil, R.; Quintana, J.B.; Knepper, T.P.; Reemtsma, T.; Berger, U. Occurrence of emerging persistent and mobile organic contaminants in European water samples. *Water Res.* **2019**, *153*, 80–90. [CrossRef] [PubMed]

35. Stuart, M.E.; Manamsa, K.; Talbot, J.C.; Crane, E.J. *Emerging Contaminants in Groundwater*; British Geological Survey Open Report; British Geological Survey: Nottingham, UK, 2011.

36. Sorensen, J.P.R.; Lapworth, D.J.; Nkhuwa, D.C.W.; Stuart, M.E.; Gooddy, D.C.; Bell, R.A.; Chirwa, M.; Kabika, J.; Liemisa, M.; Chibesa, M.; et al. Emerging contaminants in urban groundwater sources in Africa. *Water Res.* **2015**, *72*, 51–63. [CrossRef] [PubMed]

37. Lu, M.C.; Chen, Y.Y.; Chiou, M.R.; Chen, M.Y.; Fan, H.J. Occurrence and treatment efficiency of pharmaceuticals in landfill leachates. *Waste Manag.* **2016**, *55*, 257–264. [CrossRef] [PubMed]

38. Buszka, P.M.; Yeskis, D.J.; Kolpin, D.W.; Furlong, E.T.; Zaugg, S.D.; Meyer, M.T. Waste-indicator and pharmaceutical compounds in landfill-leachate-affected ground water near Elkhart, Indiana, 2000–2002. *Bull. Environ. Contam. Toxicol.* **2009**, *82*, 653–659. [CrossRef] [PubMed]

39. Kapelewska, J.; Kotowska, U.; Wiśniewska, K. Determination of personal care products and hormones in leachate and groundwater from Polish MSW landfills by ultrasound-assisted emulsification microextraction and GC-MS. *Environ. Sci. Pollut. Res.* **2016**, *23*, 1642–1652. [CrossRef] [PubMed]
40. Nowicki, Z. *Underground Waters of Provincial Cities of Poland*; Państwowy Instytut Geologiczny: Warszawa, Poland, 2007. (In Polish)

water

MDPI

Article

Assessing Surface Sediment Contamination by PBDE in a Recharge Point of Guarani Aquifer in Ribeirão Preto, Brazil

Raissa S. Ferrari [1], Alecsandra O. de Souza [1,2], Daniel L. R. Annunciação [3], Fernando F. Sodré [3] and Daniel J. Dorta [1,4,*]

[1] Faculdade de Filosofia, Ciências e Letras de Ribeirão Preto, Departamento de Química, Universidade de São Paulo, Av. Bandeirantes, 3900, Ribeirão Preto, SP 14040-901, Brazil
[2] Instituto Federal de Educação, Ciência e Tecnologia de Rondônia – Campus Porto Velho, Porto Velho, RO 76821-001, Brazil
[3] Instituto de Química, Universidade de Brasília, Brasília, DF 70910-000, Brazil
[4] National Institute for Alternative Technologies of Detection, Toxicological Evaluation and Removal of Micropollutants and Radioactives (INCT-DATREM), Unesp, Institute of Chemistry, P.O. Box 355, Araraquara, SP 14800-900, Brazil
* Correspondence: djdorta@ffclrp.usp.br

Received: 25 June 2019; Accepted: 25 July 2019; Published: 2 August 2019

check for updates

Abstract: Polybrominated diphenyl ethers (PBDEs) are used as flame retardants in several products, although they can act as neurotoxic, hepatotoxic and endocrine disruptors in organisms. In Brazil, their levels in aquatic sediments are poorly known; thus, concerns about the degree of exposure of the Brazilian population to PBDEs have grown. This study aimed to quantify the presence of PBDEs in sediment samples from an important groundwater water supply in Ribeirao Preto, Brazil, and to contribute to studies related to the presence of PBDEs in Brazilian environments. Gas chromatography coupled with Electron Capture Detection (GC-ECD) was used for quantification after submitting the samples to ultrasound-assisted extraction and clean-up steps. Results showed the presence of six PBDE, BDE-47 being the most prevalent in the samples, indicating a major contamination of the penta-PBDE commercial mixture. The concentration of ΣPBDEs (including BDE-28, -47, -66, -85, -99, -100, -138, -153, -154 and -209) varied between nd (not detected) to 5.4 ± 0.2 ng g^{-1}. Although preliminary, our data show the anthropic contamination of a direct recharge area of the Aquifer Guarani by persistent and banned substances.

Keywords: PBDE; sediment; Guarani aquifer; persistent organic pollutants; flame retardants

---

1. Introduction

Polybrominated diphenyl ethers (PBDEs) are a group of pollutants that has gained notoriety in recent decades due to their increased level in biotic and abiotic samples, not to mention several "in vitro" and "in vivo" assays that have evidenced their potential to cause damage [1–7]. PBDEs belong to the group of brominated flame retardants (BFRs) that were introduced in the 1970s as an alternative to other banned flame retardants such as polychlorinated biphenyls (PCB) and polybrominated biphenyls (PBBs) [8,9].

PBDEs are applied as a security system to retard, suppress, or inhibit the combustion process, thereby reducing risks of fires [9–11]. They are used in electronic devices like televisions and computers, circuit boards, cables, automotive components, and construction materials; they are also employed in the textile industry [12,13]. The chemical structure of PBDEs comprises of two phenyl rings linked by an ether bond, which enables up to 10 substitutions with bromine and leads to up 209 congeners with distinct physicochemical characteristics, depending on the order of chemical bromine substitution [2,14]. These compounds are also considered harmful persistent organic pollutants because they are little soluble in water, have high octanol-water partition coefficient (K_{ow}), can persist in the environment, can accumulate in living organisms [15], and have shown toxicity to several organisms.

Dispersion of these compounds in the environment is aggravated by their weak chemical interaction with the surface of polymers, which facilitates their removal from manufactured products as well as their volatilization and dust formation during the use of treated products [8]. Their lipophilicity and persistence contribute to their accumulation in biota, soil, air particles, sewage, aquatic particles, sediments, and food [12,16]. PBDEs have been detected even in polar regions due to transport mechanisms in different trophic systems [17–19].

Once released into the environment, PBDEs can accumulate in house dust and food, and they can be biomagnified along the food chain [20–24]. Therefore, humans may be exposed to PBDEs when they inhale dust [22,25] and/or ingest contaminated food. The absorption of PBDEs is worrisome because these compounds are endocrine disrupting chemicals [26,27] and can thus lead to several diseases of the endocrine system [28,29]. There have also been evidences of their neurotoxicity [30–32] and hepatoxicity [5–7]. The toxic effects of PBDEs to humans and other organisms are related to the original substances and also to their metabolites, such as OH-PBDEs, that are formed by their metabolization by cytochrome P450s enzymes [33–35].

Levels of PBDEs ranging from 0.04 ng g^{-1} to 527,000 ng g^{-1} have been found in sediment and dust [36–40]. Such levels depend on factors that include local anthropogenic activities, presence of degrading microorganisms, and environmental photodegradation [41–43].

The use of PBDEs has been restricted and/or prohibited in parts of the world, such as Europe Union, United States of America, Japan and Australia and some congeners banished by the Stockholm Convention on persistent organic pollutants (POPs), a convention aiming to protect human health and the environment from POPs. However, there are no restrictions on the use of PBDEs in other locations, such as Brazil, even though the country has already ratified its signature to the convention, which has culminated to exposure to unknown concentrations of PBDEs in different regions [44]. Currently, knowledge of the levels of exposure to PBDEs in Brazil are limited because there are few reports about their presence in abiotic and biotic samples [2,11,45].

In Brazil, some sites are interesting targets to evaluate the level of environmental contamination with PBDEs. For example, the city of Ribeirao Preto in Brazil is located on a recharge area of the Guarani Aquifer, a groundwater reservoir that supplies all the drinking water to the city [46] However, this area is vulnerable to pollution due to the presence of strong agro-industrial activities in the region that are able to release many toxic substances into the environment, consequently affecting the quality of water and sediments [47–50]. In this context, the presence of PBDEs in sediments from the Saibro Lagoon, a recharge point of the Guarani Aquifer in the city of Ribeirao Preto, is a matter of concern and the object of the present investigation.

2. Experimental

2.1. Chemicals and Reagents

All the reagents were of analytical grade or higher purity. Acetone, *n*-pentane, and isooctane were pesticide grade, acquired from Tedia (Fairfield, CT, USA). Anhydrous sodium sulfate and silica gel 60 (0.05 to 0.2 mm) were purchased from Vetec (Rio de Janeiro, Brazil). The Lake Michigan Study standard (BDE-LMS) containing BDE-28, -47, -66, -85, -99, -100, -138, -153, and -154 was purchased from AccuStandard (New Haven, CT, USA). A 10 µg mL^{-1} stock solution, prepared in isooctane was used to prepare working solutions (0.01, 0.1, 1.0, 5.0, 10.0, and 20.0 ng mL^{-1}) for the analytical curves. The 50 µg mL^{-1} BDE-209 standard prepared in isooctane/toluene (9:1 v/v) was also obtained from AccuStandard and used to prepare appropriate working solutions. Quantification was carried out by external calibration curves. Figure 1 shows the structure of the 10 PBDEs investigated in this work that are also among the main congeners investigated in environment and biological fluids.

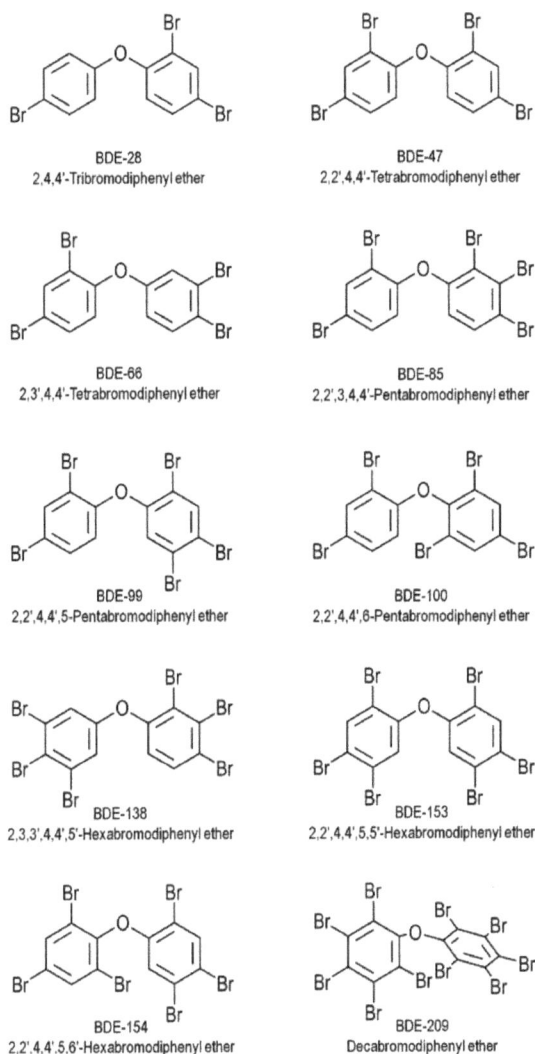

BDE-28
2,4,4'-Tribromodiphenyl ether

BDE-47
2,2',4,4'-Tetrabromodiphenyl ether

BDE-66
2,3',4,4'-Tetrabromodiphenyl ether

BDE-85
2,2',3,4,4'-Pentabromodiphenyl ether

BDE-99
2,2',4,4',5-Pentabromodiphenyl ether

BDE-100
2,2',4,4',6-Pentabromodiphenyl ether

BDE-138
2,3,3',4,4',5'-Hexabromodiphenyl ether

BDE-153
2,2',4,4',5,5'-Hexabromodiphenyl ether

BDE-154
2,2',4,4',5,6'-Hexabromodiphenyl ether

BDE-209
Decabromodiphenyl ether

Figure 1. Structures of the 10 polybrominated diphenyl ethers (PBDEs) investigated in this work.

2.2. Study Site and Sampling

This work was carried out in the city of Ribeirao Preto, located in the northeastern portion of the State of Sao Paulo, Brazil. It is typically urban, and its activities are focused on trade, services, and sugarcane culture. Its climate is tropical humid, but the weather is dry in winter, when we collected the samples. Samples of surface sediment of Saibro Lagoon were collected in accessible points as can be seen in Figure 2.

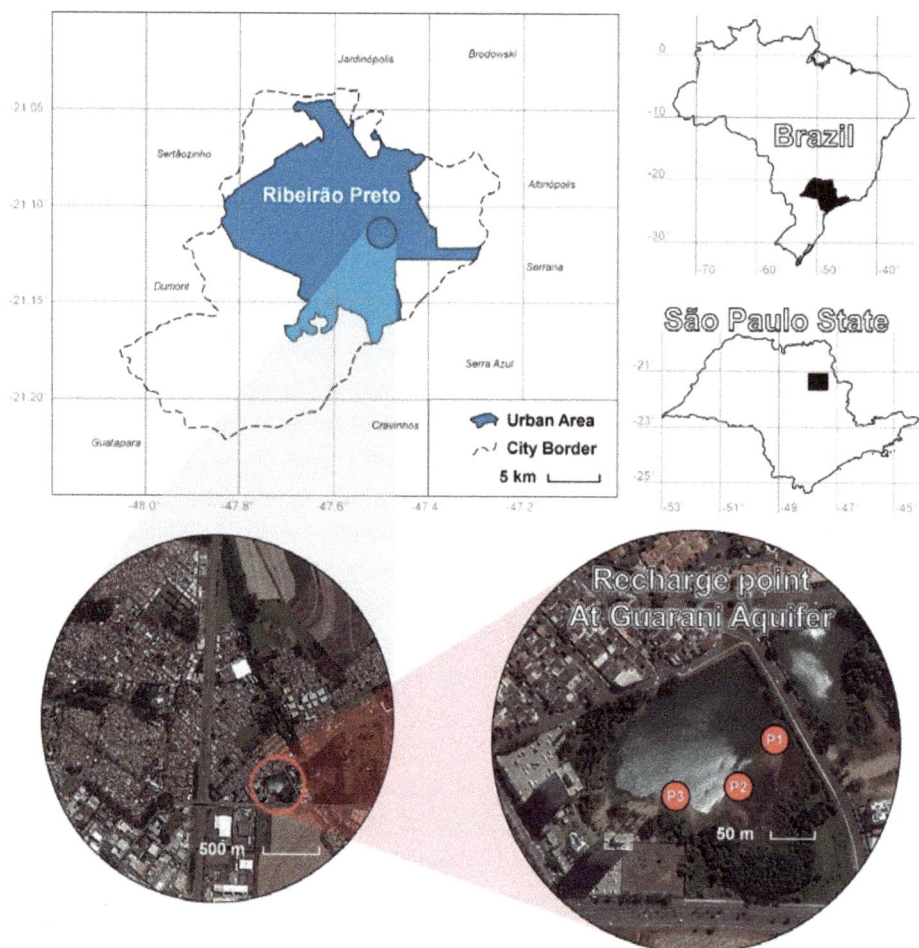

Figure 2. Map of the sampling site. In the bottom left, the red circle indicates the location of sampling points in the Saibro Lagoon.

Samples were collected in June 2014 using a cylindrical corer. The obtained sediment pellet was packaged in clean amber glass bottles and stored in a refrigerator at a temperature below 4 °C until further preparation steps. Dry sediments were obtained by transferring portions of the samples to Petri dishes that were capped individually with aluminum foil to prevent cross contamination. Then, small holes were made in the cover foil to allow water to evaporate under room temperature (25 °C). Petri dishes were set in an exhaust hood and allowed to rest for 48 h.

2.3. PBDE Extraction and Quantification

The sediment was analyzed according to a method described and validated by Annunciação et al. [51]. Briefly, 3000 g of dried sediments and 10 mL of the extraction solvent mixture (acetone/*n*-pentane, 1:1 v/v) were transferred to glass tubes. Tubes were sealed, agitated for 30 s and suspensions were sonicated in a Cole-Parmer 8893 bath (Vernon Hills, IL, USA) operating at 40 kHz for 5 min at 25 °C and centrifuged (Kindly, KC5, São Paulo, Brazil) for 5 min at 1000× *g*. Then, the supernatant was transferred to a 250 mL flat-bottomed flask. The remaining solid fraction was submitted to ultrasonic extraction four more times, to produce a final composite extract. Isooctane was added as solvent keeper (1.0 mL), and the extract volume was reduced to approximately 2 mL in a rotary evaporator (Fisatom 801, São Paulo, Brazil). Interferences were minimized by eluting the extract with 50 mL of *n*-pentane through a 30-cm column filled with acid, basic, and neutral silica gel (3 g, 2 g, and 3 g, respectively) as well as with anhydrous sodium sulfate between the silica gel phases. Details on the preparation of the silicas are available elsewhere [51]. After re-concentration to 1.0 mL of the keeper, the extracts were treated with copper strips to avoid interference of sulfur species during the quantification of the analytes. Finally, in order to improve detectability, extracts volume was reduced to 0.25 mL in a gently flow of N_2.

The content of PBDEs in the final extracts was assessed by gas chromatography with an electron capture detector (Shimadzu GC 2010 Plus, Kyoto, Japan). The PBDEs were separated with a Zebron ZB-XLB-HT (15 m × 0.25 mm × 0.25 μm, Agilent Technologies, Santa Clara, USA) chromatographic column for BDE-209 and with a SLM TM-5ms (30 m × 0.25 mm × 0.25 μm, Supelco, Bellefonte, PA, USA) column for the less brominated congeners. The PBDEs were quantified by external calibration in triplicate. Limits of detection (LOD) were calculated by the signal-to-noise approach where ratios of three corresponded to LOD. Table 1 summarizes selected analytical parameters of the method.

Table 1. Analytical parameters for the determination of the selected PBDE.

Congeners	Retention time/min	LOD/(ng g^{-1}) [b]	LOQ/(ng g^{-1}) [c]	Recovery/% [d]
BDE-28	28.1	0.14	0.48	98.5 ± 2.5
BDE-47	33.1	0.15	0.52	98.2 ± 4.3
BDE-66	34.1	0.17	0.57	97.8 ± 4.7
BDE-85	40.4	0.17	0.58	92.9 ± 6.0
BDE-99	38.0	0.2	0.67	97.9 ± 6.0
BDE-100	36.9	0.17	0.57	97.6 ± 6.5
BDE-138	45.6	0.24	0.80	89.8 ± 2.8
BDE-153	43.0	0.21	0.72	90.5 ± 4.5
BDE-154	41.4	0.17	0.57	92.4 ± 7.1
BDE-209	23.5 [a]	0.14	0.48	94.2 ± 5.3

[a] In a single component analysis using a 15 m chromatographic column, [b] method limit of detection, [c] method limit of quantification, [d] for a fortified (5.0 ng g^{-1}) blank sediment (n = 7).

It is important to point out that the analytical method was previously validated [51] using fortified blank sediments and a certified reference sediment (RTC-SQC072) supplied by RTC (Laramie, WY, USA) containing PCBs. In this case, intraday recoveries (*n* = 7) varied from 89.8 ± 2.8% (BDE-138) to 98.5 ± 2.5% (BDE-28), as shown in Table 1, and inter-day recoveries (*n* = 9) varied between 87.7 ± 6.6% (BDE-138) and 98.7 ± 1.4% (BDE-28).

3. Results and Discussion

Figure 3 shows the chromatograms obtained during the determination of the less brominated congeners in the sediment samples.

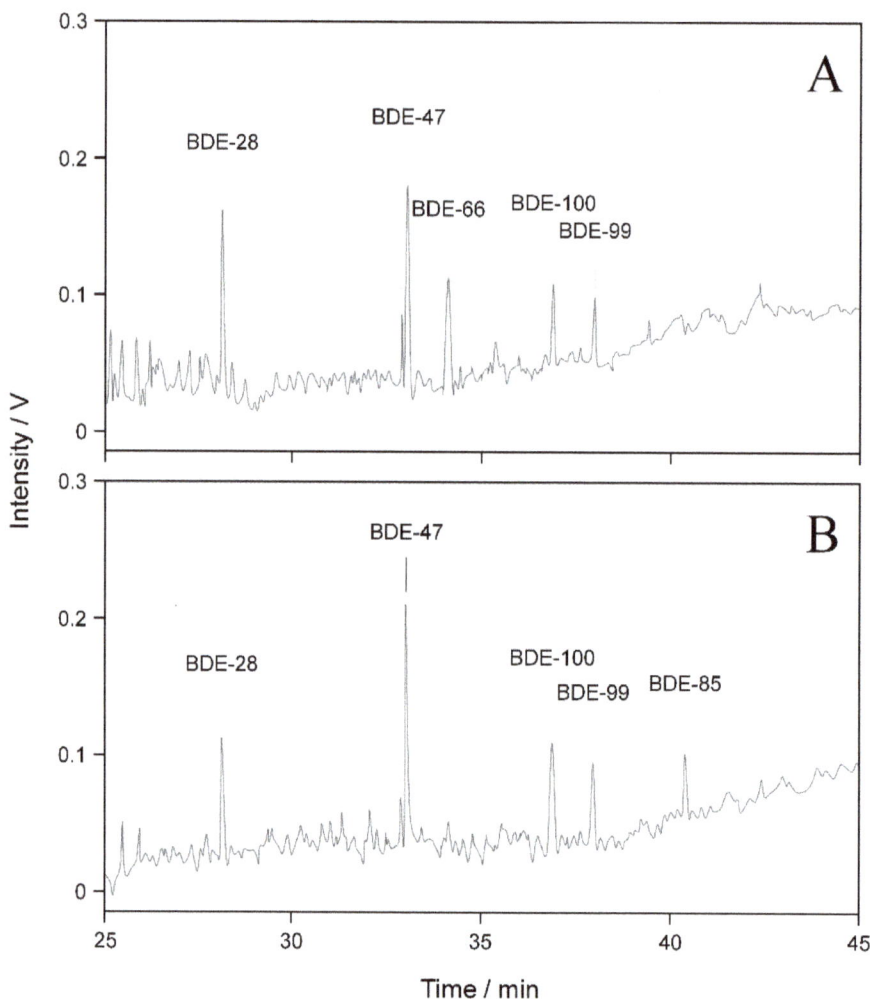

Figure 3. Chromatograms of the sediment extracts ((**A**) P2 sample; (**B**) P3 sample) obtained during determination of the less brominated PBDEs.

In Figure 3 it is possible to observe that PBDE were detected only in the samples collected in Points 2 and 3. The sample collected at Point 1 did not present detectable concentrations of PBDEs probably due to the texture of the collected material, which consisted mostly of sand with naturally low specific surface area. Also, during the clean-up step, it was possible to notice that this sample presented small quantities of organic carbon, which may contribute to the fixation of PBDE in the sediment matrix. Furthermore, it was already shown that both the size of the grains and the content of organic matter can influence the absorption and accumulation of PBDEs in the sediments [52].

Chromatograms for the sediment samples from Points 2 and 3 revealed distinct analytical signals for the investigated PBDEs. However, in both cases, it is possible to notice that peaks height were near the chromatographic noise, revealing that low levels of PBDE were detected in the samples. Chromatograms did not reveal the presence of BDE-209 during the single component analysis. Table 2 lists the concentration of the investigated PBDEs in the sediment samples.

Table 2. Concentrations (ng g^{-1} dry weight) of the investigated congeners in the sediment samples collected at the Saibro Lagoon.

Congeners	Point 1	Point 2	Point 3
BDE-28	nd	0.68 ± 0.04	0.58 ± 0.03
BDE-47	nd	2.10 ± 0.06	2.7 ± 0.1
BDE-66	nd	0.76 ± 0.03	nd
BDE-85	nd	0.24 ± 0.02	0.80 ± 0.05
BDE-99	nd	0.25 ± 0.02	0.94 ± 0.08
BDE-100	nd	Nd	0.35 ± 0.02
BDE-138	nd	Nd	nd
BDE-153	nd	Nd	nd
BDE-154	nd	Nd	nd
BDE-209	nd	Nd	nd
Σ_{10}PBDE	-	4.03 ± 0.09	5.4 ± 0.2

nd—not detected.

Six of the 10 investigated congeners were detected in the sediment samples, i.e., BDE-28, -47, -66, -85, -99, and -100. The concentration of PBDEs in the Point 2 sample varied between 0.24 ± 0.02 (BDE-85) and 2.10 ± 0.06 ng g^{-1} (BDE-47). For the sample collected at Point 3, the concentrations of PBDEs ranged from 0.35 ± 0.02 (BDE-100) to 2.7 ± 0.1 ng g^{-1} (BDE-47). In both samples, BDE-47 was the most prevalent congener indicating a major contamination of the penta-PBDE commercial mixture [53,54]. This same conclusion was made by Annunciação et al. [51] investigating samples from the Paranoá Lake in the Brazilian Federal District. Both results reveal an important contribution of the penta-BDE mixture in the consumer goods used in Brazil.

In general, BDE-47, -99, -100, -153, -154, and -183 were the most commonly identified congeners because they are the major compounds in the penta-BDE commercial mixture [55]. The congener BDE-47 is most often employed in the furniture and upholstery industry [56]. The mixture penta-BDE, which includes BDE-47, -99, -100, -153, and -154, is typically applied in furniture, whereas the octa- and deca-BDEs mixtures are used in the manufacture of many types of polymers, especially those used in televisions, computers, and cables [8,56].

The total concentration of PBDEs (expressed as Σ_{10}PBDE) was higher in the sample obtained at Point 3 (5.4 ± 0.2 ng g^{-1}) as compared to the value calculated for Point 2 (4.03 ± 0.09 ng g^{-1}). In both cases, BDE-47 accounted for more than 50% of the total concentration of PBDEs, which showed that it was the most prevalent congener in the samples, followed by BDE-66 and BDE-99 in the samples extracted at Points 2 and Point 3, respectively.

According to the literature, BDE-47 and -99 are the PBDE congeners that are most frequently detected in environmental samples and biological fluids [57–60]. These congeners have also been identified in Brazilian biological samples reported by Dorneles and co-workers [61], who verified methoxylated metabolites of BDE-47 and BDE-68 in cetaceans of the Brazilian marine coast. BDE-47 and BDE-85 have also been detected in dolphins in the Paraiba do Sul River in Brazil [62], and the presence of PBDE commercial mixtures in whitemouth croakers from southeastern Brazil [45]. These data show main accumulation of BDE-47 in environmental samples [63,64].

The presence of PBDEs in the investigated region is explained by the habits of the surrounding population regarding the disposal of household waste. Weeks after the sampling takes place, the municipal government of Ribeirão Preto reported that they removed half a ton of garbage, mostly furniture and electronics, from the lake [65]. Although contamination of the environment may have occurred by the disposal of municipal waste, it is important to consider other important routes of urban water contamination by PBDE. In this context, much evidence indicates that PBDEs reach aquatic environments by dumping sewage, whether raw or treated [66–68]. In this case, as described by Annunciação et al. [11], the presence of PBDE in urban wastewater can be influenced by a sequence of processes that begin with the cleaning of surfaces in domestic, commercial or industrial environments.

Thus, dust particles are transferred to the washing waters and disposed either through drains, reaching storm water or sewage systems, or through sewers, reaching rainwater drainage networks. Furthermore, PBDEs can remain in the atmosphere in the form of particles that can precipitate with the rain, being further deposited in sediments, where they accumulate [69,70].

Our data represent one of the first involving the contamination of Brazilian aquatic sediments by PBDE. Annunciação et al. [51] investigated the presence of nine congeners in samples from Paranoá Lake, Brazil, collected in the proximity of two sewage treatment plants and evidenced the prevalence of BDE -47 and -66. Other Brazilian research groups have also evidenced environmental contamination of PBDE as they were able to quantify different congeners in samples from fish and marine mammals on the coast of the state of Rio de Janeiro, Brazil, as pointed out earlier [61,71].

In order to better visualize the results obtained in the present work, we compared the concentrations of PBDEs found herein with the concentrations reported worldwide, as depicted in Table 3.

Table 3. Concentrations (ng g^{-1} dry weight) of PBDEs measured in this study in comparison with those reported form aquatic sediments collected around the world.

Country	Aquatic System	n [a]	∑PBDEs [b]	References
Brazil	Saibro Lagoon	10	nd [c]–5.4	This Study
Brazil	Paranoá Lake	9	2.5–8.1	[51]
Chile	Copncepción Bay	10	0.02–21	[72]
China	Baiyangdian Lake	8	0.05–5.03	[73]
China	Fuhe River	8	0.13–6.39	[73]
China	Shanghai rivers	16	0.44–12.0	[74]
China	Jiaojiang River	10	8.93–45.0	[42]
South Korea	Shihwa Lake	23	1.13–18700	[39]
Canada	Nigara River	9	1.10–148	[75]
USA	White Lake	23	0.39–2,4	[76]
USA	Muskegon Lake	23	0.98–3.9	[76]
Italy	Maggiore Lake	8	0.02–27.1	[70]
Russia	Olkhon Island	40	0.164–0.670	[69]

[a] Number of PBDEs congeners analyzed in sample. [b] Sum of all target PBDE congeners except for BDE 209. [c] Not detected.

Compared to some regions in the Fuhe River, the Baiyangdian Lake, and the Chaohu Lake, all of them in China, and on the Olkhon Island, in Russia, the mean concentrations of ∑PBDEs without BDE-209 detected in this study were higher. Cheng-Yu, an economic region in China, presented much lower mean values of PBDEs as compared to the Saibro Lagoon investigated herein. Compared to some regions in Korea (four major rivers) or even in China (Jiaojian River), our values were much lower, which means that this lagoon in Ribeirão Preto was moderately contaminated with PBDEs.

The toxicity of PBDEs to living beings will also depend on the rate of absorption and on the metabolic system of the exposed species. Nevertheless, we must keep in mind that exposure to these pollutants or biomagnification through the food chain cause PBDEs to accumulate in lipids. For example, Hites [63] has shown that the concentration of PBDEs in Canadian Arctic seals has increased throughout the years, even though the Canadian Arctic is a remote region without direct release of these compounds into the environment. The increased presence of this compound can be related to its efficient atmospheric transport and bioaccumulation [77].

4. Conclusions

PBDE congeners were found in samples collected from the Saibro Laggon (Ribeirão Preto, Brazil), in a recharge point of Guarani Aquifer. The highest Σ_{10}PBDE concentration, i.e., 5.4 ± 0.2 ng g^{-1}, agreed with the lowest levels determined in sediments collected around the world. BDE-47 was the most prevalent congener, accounting for more than 50% of the PBDE concentration in the samples evidencing the major use of the penta-PBDE commercial mixture widely used as flame retardant in furniture. Contamination of the aquifer recharge area probably occurs via inappropriate disposal of municipal solid wastes. Our results have shown that this important recharge area of is already contaminated with PBDEs, which may affect local population due to their potential to bioaccumulate and to induce damage in all exposed organisms. This contamination can still increase in the coming years considering that the use of PBDE is not controlled in Brazil. As the preservation of this site is vital, this work will help the scientific community to clarify the degree of exposure to PBDEs in Brazil.

Author Contributions: Conceptualization, D.J.D.; Data curation, R.S.F., A.O.d.S. and D.L.R.A.; Formal analysis, D.L.R.A.; Funding acquisition, D.J.D.; Investigation, R.S.F. and A.O.d.S.; Methodology, D.L.R.A.; Project administration, D.J.D.; Resources, D.J.D.; Supervision, F.F.S.; Validation, D.L.R.A.; Writing—original draft, R.S.F. and A.O.d.S.; Writing—review & editing, F.F.S. and D.J.D.

Funding: This study was financed in part by the Coordenação de Aperfeiçoamento de Pessoal de Nível Superior—Brasil (CAPES)—Finance Code 001. This work was partially supported by the National Institute for Advanced Analytical Science and Technology (INCTAA), project CNPq 573894/2008-6. Dorta is also thankful for support from Conselho Nacional de Desenvolvimento Científico e Tecnológico (CNPq): project 308659/2012-0.

Acknowledgments: At the University of Brasilia, the authors also thanks the Analytical Center of the Institute of Chemistry (CAIQ) for providing the analytical equipment used in this work.

Conflicts of Interest: The authors declare no conflict of interest. The funders had no role in the design of the study; in the collection, analyses, or interpretation of data; in the writing of the manuscript, or in the decision to publish the results.

References

1. Kalantzi, O.I.; Martin, F.L.; Thomas, G.O.; Alcock, R.E.; Tang, H.R.; Drury, S.C.; Carmichael, P.L.; Nicholson, J.K.; Jones, K.C. Different levels of polybrominated diphenyl ethers (PBDEs) and chlorinated compounds in breast milk from two U.K. *Environ. Health Perspect.* **2004**, *112*, 1085. [CrossRef] [PubMed]
2. Rodrigues, E.M.; Ramos, A.B.A.; Cabrini, T.M.B.; Fernandez, M.A.; Dos, S. The occurrence of polybrominated diphenyl ethers in Brazil: A review. *Int. J. Environ. Heal.* **2015**, *7*, 247. [CrossRef]
3. Stoker, T.E.; Cooper, R.L.; Lambright, C.S.; Wilson, V.S.; Furr, J.; Gray, L.E. In vivo and in vitro anti-androgenic effects of DE-71, a commercial polybrominated diphenyl ether (PBDE) mixture. *Toxicol. Appl. Pharmacol.* **2005**, *207*, 78. [CrossRef]
4. Xing, T.; Chen, L.; Tao, Y.; Wang, M.; Chen, J.; Ruan, D.-Y. Effects of decabrominated diphenyl ether (PBDE 209) exposure at different developmental periods on synaptic plasticity in the dentate gyrus of adult rats in vivo. *Toxicol. Sci.* **2009**, *110*, 401. [CrossRef] [PubMed]
5. Dorta, D.J.; de Souza, A.O.; Pereira, L.C.; Pazin, M.; de Oliveira, G.A.R.; de Oliveira, D.P. *Polybrominated Diphenyl Ether (PBDE) Flame Retardant as Emergent Environmental Pollutants: An Overview on Their Environmental Contamination and Toxicological Properties. Advances in Environmental Research*; Daniels, J.A., Ed.; Nova Science Publishers: New York, NY, USA, 2013; Volume 28, pp. 43–73.
6. Pazin, M.; Pereira, L.C.; Dorta, D.J. Toxicity of brominated flame retardants, BDE-47 and BDE-99 stems from impaired mitochondrial bioenergetics. *Toxicol. Mech. Methods* **2015**, *25*, 34. [CrossRef]
7. Souza, A.O.; Pereira, L.C.; Oliveira, D.P.; Dorta, D.J. BDE-99 congener induces cell death by apoptosis of human hepatoblastoma cell line - HepG2. *Toxicol. In Vitro* **2013**, *27*, 580. [CrossRef]
8. Król, S.; Zabiegała, B.; Namieśnik, J. PBDEs in environmental samples: Sampling and analysis. *Talanta* **2012**, *93*, 1. [CrossRef]
9. Pieroni, M.C.; Leonel, J.; Fillmann, G. Retardantes de chama bromados: Uma revisão. *Quim. Nova* **2017**, *40*, 317. [CrossRef]
10. Daubié, S.; Bisson, J.-F.; Lalonde, R.; Schroeder, H.; Rychen, G. Neurobehavioral and physiological effects of low doses of polybrominated diphenyl ether (PBDE)-99 in male adult rats. *Toxicol. Lett.* **2011**, *204*, 57. [CrossRef]
11. Annunciação, D.L.R.; Almeida, F.V.; Hara, E.L.Y.; Grassi, M.T.; Sodré, F.F. Éteres difenílicos polibromados (PBDE) como contaminantes persistentes: Ocorrência, comportamento no ambiente e estratégias analíticas. *Quim. Nova* **2018**, *41*, 782. [CrossRef]
12. Hyötyläinen, T.; Hartonen, K. Determination of brominated flame retardants in environmental samples. *TrAC Trends Anal. Chem.* **2002**, *21*, 13. [CrossRef]

13. Segev, O.; Kushmaro, A.; Brenner, A. Environmental Impact of Flame Retardants (Persistence and Biodegradability). *Int. J. Environ. Res. Public Health* **2009**, *6*, 478–491. [CrossRef]
14. Frouin, H.; Lebeuf, M.; Hammill, M.; Masson, S.; Fournier, M. Effects of individual polybrominated diphenyl ether (PBDE) congeners on harbour seal immune cells in vitro. *Mar. Pollut. Bull.* **2010**, *60*, 291. [CrossRef]
15. Sapozhnikova, Y.; Lehotay, S.J. Multi-class, multi-residue analysis of pesticides, polychlorinated biphenyls, polycyclic aromatic hydrocarbons, polybrominated diphenyl ethers and novel flame retardants in fish using fast, low-pressure gas chromatography–tandem mass spectrometry. *Anal. Chim. Acta* **2013**, *758*, 80. [CrossRef]
16. Birnbaum, L.S.; Staskal, D.F. Brominated flame retardants: Cause for concern? *Environ. Health Perspect.* **2004**, *112*, 9. [CrossRef]
17. Baek, S.-Y.; Choi, S.-D.; Chang, Y.-S. Three-Year Atmospheric Monitoring of Organochlorine Pesticides and Polychlorinated Biphenyls in Polar Regions and the South Pacific. *Environ. Sci. Technol.* **2011**, *45*, 4475. [CrossRef]
18. Dietz, R.; Rigét, F.F.; Sonne, C.; Letcher, R.J.; Backus, S.; Born, E.W.; Kirkegaard, M.; Muir, D.C.G. Age and seasonal variability of polybrominated diphenyl ethers in free-ranging East Greenland polar bears (Ursus maritimus). *Environ. Pollut.* **2007**, *146*, 166. [CrossRef]
19. McKinney, M.A.; Letcher, R.J.; Aars, J.; Born, E.W.; Branigan, M.; Dietz, R.; Evans, T.J.; Gabrielsen, G.W.; Peacock, E.; Sonne, C. Flame retardants and legacy contaminants in polar bears from Alaska, Canada, East Greenland and Svalbard, 2005–2008. *Environ. Int.* **2011**, *37*, 365. [CrossRef]
20. Allen, J.G.; McClean, M.D.; Stapleton, H.M.; Webster, T.F. Critical factors in assessing exposure to PBDEs via house dust. *Environ. Int.* **2008**, *34*, 1085. [CrossRef]
21. Jeong, Y.; Lee, S.; Kim, S.; Choi, S.-D.; Park, J.; Kim, H.-J.; Lee, J.J.; Choi, G.; Choi, S.; Kim, S.; et al. Infant exposure to polybrominated diphenyl ethers (PBDEs) via consumption of homemade baby food in Korea. *Environ. Res.* **2014**, *134*, 396. [CrossRef]
22. Jones-Otazo, H.A.; Clarke, J.P.; Diamond, M.L.; Archbold, J.A.; Ferguson, G.; Harner, T.; Richardson, G.M.; Ryan, J.J.; Wilford, B. Is House Dust the Missing Exposure Pathway for PBDEs. *Environ. Sci. Technol.* **2005**, *39*, 5121. [CrossRef]
23. Wang, W.; Zheng, J.; Chan, C.Y.; Huang, M.J.; Cheung, K.C.; Wong, M.H. Health risk assessment of exposure to polybrominated diphenyl ethers (PBDEs) contained in residential air particulate and dust in Guangzhou and Hong Kong. *Atmos. Environ.* **2014**, *89*, 786. [CrossRef]
24. Wu, N.; Herrmann, T.; Paepke, O.; Tickner, J.; Hale, R.; Harvey, E.; La Guardia, M.; McClean, M.D.; Webster, T.F. Human exposure to PBDEs: Associations of PBDE body burdens with food consumption and house dust concentrations. *Environ. Sci. Technol.* **2007**, *41*, 1584. [CrossRef]
25. Wilford, B.H.; Shoeib, M.; Harner, T.; Zhu, J.; Jones, K.C. Polybrominated diphenyl ethers in indoor dust in ottawa, canada: Implications for sources and exposure. *Environ. Sci. Technol.* **2005**, *39*, 7027. [CrossRef]
26. Ellis-Hutchings, R.G.; Cherr, G.N.; Hanna, L.A.; Keen, C.L. Polybrominateddiphenyl ether (PBDE)-induced alterations in vitamin A and thyroid hormoneconcentrations in the rat during lactation and early postnatal development. *Toxicol. Appl. Pharmacol.* **2006**, *215*, 135. [CrossRef]
27. Yu, L.; Lam, J.C.W.; Guo, Y.; Wu, R.S.S.; Lam, P.K.S.; Zhou, B. Parental Transfer of Polybrominated Diphenyl Ethers (PBDEs) and Thyroid Endocrine Disruption in Zebrafish. *Environ. Sci. Technol.* **2011**, *45*, 10652. [CrossRef]
28. Hauser, R.; Skakkebaek, N.E.; Hass, U.; Toppari, J.; Juul, A.; Andersson, A.M.; Kortenkamp, A.; Heindel, J.J.; Trasande, L. Male reproductive disorders, diseases, and costs of exposure to endocrine-disrupting chemicals in the European Union. *J. Clin. Endocrinol. Metab.* **2015**, *100*, 1267. [CrossRef]
29. Oulhote, Y.; Chevrier, J.; Bouchard, M.F. Exposure to Polybrominated Diphenyl Ethers (PBDEs) and Hypothyroidism in Canadian Women. *J. Clin. Endocrinol. Metab.* **2016**, *101*, 590. [CrossRef]
30. Chen, J.; Liufu, C.; Sun, W.; Sun, X.; Chen, D. Assessment of the neurotoxic mechanisms of decabrominated diphenyl ether (PBDE-209) in primary cultured neonatal rat hippocampal neurons includes alterations in second messenger signaling and oxidative stress. *Toxicol. Lett.* **2010**, *192*, 431. [CrossRef]
31. Costa, L.G.; Giordano, G. Is decabromodiphenyl ether (BDE-209) a developmental neurotoxicant? *Neurotoxicology* **2011**, *32*, 9. [CrossRef]
32. Costa, L.G.; Giordano, G. Developmental neurotoxicity of polybrominated diphenyl ether (PBDE) flame retardants. *Neurotoxicology* **2007**, *28*, 1047. [CrossRef]

33. Dong, H.; Li, Z.; Man, X.; Zhou, J.; Lu, H.; Wang, S. Identification of the metabolites of polybrominated diphenyl ether 99 and its related cytochrome P450s. *J. Biomed. Res.* **2010**, *24*, 223. [CrossRef]
34. Zhang, B.; Xu, T.; Huang, G.; Yin, D.; Zhang, Q.; Yang, Y. Neurobehavioral effects of two metabolites of BDE-47 (6-OH-BDE-47 and 6-MeO-BDE-47) on zebrafish larvae. *Chemosphere* **2018**, *200*, 30. [CrossRef]
35. Gross, M.S.; Butryn, D.M.; McGarrigle, B.P.; Aga, D.S.; Olson, J.R. Primary Role of Cytochrome P450 2B6 in the Oxidative Metabolism of 2,2′,4,4′,6-Pentabromodiphenyl Ether (BDE-100) to Hydroxylated BDEs. *Chem. Res. Toxicol.* **2015**, *28*, 672. [CrossRef]
36. Besis, A.; Samara, C. Polybrominated diphenyl ethers (PBDEs) in the indoor and outdoor environments e A review on occurrence and human exposure. *Environ. Pollut.* **2012**, *169*, 217. [CrossRef]
37. Huang, Y.; Chen, L.; Peng, X.; Xu, Z.; Ye, Z. PBDEs in indoor dust in South-Central China: Characteristics and implications. *Chemosphere* **2010**, *78*, 169. [CrossRef]
38. Kang, Y.; Wang, H.S.; Cheung, K.C.; Wong, M.H. Polybrominated diphenyl ethers (PBDEs) in indoor dust and human hair. *Atmos. Environ.* **2011**, *45*, 2386. [CrossRef]
39. Moon, H.B.; Choi, M.; Yu, J.; Jung, R.H.; Choi, H.G. Contamination and potential sources of polybrominated diphenyl ethers (PBDEs) in water and sediment from the artificial Lake Shihwa, Korea. *Chemosphere* **2012**, *88*, 837. [CrossRef]
40. Tan, J.; Cheng, S.M.; Loganath, A.; Chong, Y.S.; Obbard, J.P. Polybrominated diphenyl ethers in house dust in Singapore. *Cheimosphere* **2007**, *66*, 985. [CrossRef]
41. Shi, S.-X.; Huang, Y.-R.; Zhou, L.; Zhang, L.-F.; Dong, L.; Yang, W.-L.; Zhang, X.-L. Changes of polybrominated diphenyl ethers and polychlorinated biphenyls in surface soils from urban agglomeration of the Yangtze River Delta, in China between 2003 and 2012. *Environ. Sci. Pollut. Res.* **2015**, *22*, 9766. [CrossRef]
42. Yang, S.; Fu, Q.; Teng, M.; Yang, J. Polybrominated diphenyl ethers (PBDEs) in sediment and fish tissues from Lake Chaohu, central eastern China. *Arch. Environ. Prot.* **2015**, *41*, 12. [CrossRef]
43. Wang, Z.; Na, G.; Ma, X.; Ge, L.; Lin, Z.; Yao, Z. Characterizing the distribution of selected PBDEs in soil, moss and reindeer dung at Ny-Ålesund of the Arctic. *Chemosphere* **2015**, *137*, 9. [CrossRef]
44. Makey, C.M.; McClean, M.D.; Sjödin, A.; Weinberg, J.; Carignan, C.C.; Webster, T.F. Temporal variability of polybrominated diphenyl ether (PBDE) serum concentrations over one year. *Environ. Sci. Technol.* **2014**, *48*, 14642. [CrossRef]
45. Pizzochero, A.C.; De La Torre, A.; Sanz, P.; Navarro, I.; Michel, L.N.; Lepoint, G.; Das, K.; Shnitzler, J.G.; Chenery, S.R.; McCarthy, I.D.; et al. Occurrence of legacy and emerging organic pollutants in whitemouth croakers from Southeastern Brazil. *Sci. Total Environ.* **2019**, *682*, 719. [CrossRef]
46. DAERP (Departamento de Água e Esgoto de Ribeirão Preto) – Água. Available online: http://www.ribeiraopreto.sp.gov.br/daerp/i04agua.php (accessed on 20 May 2018).
47. Conceição, F.T.; Cunha, R.; Sardinha, D.S.; Souza, A.D.G.; Sinelli, O. Hidrogeoquímica do aqüífero guarani na área urbana de ribeirão preto (SP). *Geociencias* **2009**, *28*, 65.
48. Rabelo, J.L.; Wendland, E. Assessment of groundwater recharge and water fluxes of the Guarani Aquifer System, Brazil. *Hydrogeol. J.* **2009**, *17*, 1733. [CrossRef]
49. Scanlon, B.R.; Healy, R.W.; Cook, P.G. Choosing appropriate techniques for quantifying groundwater recharge. *Hydrogeol. J.* **2002**, *10*, 18. [CrossRef]
50. Villar, P.C.; Ribeiro, W.C. Sociedade e gestão do risco: O aquífero Guarani em Ribeirão Preto-SP, Brasil. *Rev. Geogr. Norte Gd.* **2009**, *43*, 51. [CrossRef]
51. Annunciação, D.L.R.; Almeida, F.V.; Sodré, F.F. Method development and validation for the determination of polybrominated diphenyl ether congeners in Brazilian aquatic sediments. *Microchem. J.* **2017**, *133*, 43.
52. Dinn, P.M.; Johannessen, S.C.; Ross, P.S.; Macdonald, R.W.; Whiticar, M.J.; Lowe, C.J.; van Roodselaar, A. PBDE and PCB accumulation in benthos near marine wastewater outfalls: The role of sediment organic carbon. *Environ. Pollut.* **2012**, *171*, 241. [CrossRef]
53. Björklund, J.A.; Thuresson, K.; Cousins, A.P.; Sellström, U.; Emenius, G.; De Wit, C.A. Indoor Air Is a Significant Source of Tri-decabrominated Diphenyl Ethers to Outdoor Air via Ventilation Systems. *Environ. Sci. Technol.* **2012**, *46*, 5876. [CrossRef]
54. Harrad, S.; Hazrati, S.; Ibarra, C. Concentrations of Polychlorinated Biphenyls in Indoor Air and Polybrominated Diphenyl Ethers in Indoor Air and Dust in Birmingham, United Kingdom: Implications for Human Exposure. *Environ. Sci. Technol.* **2006**, *40*, 4633. [CrossRef]

55. Darnerud, P.O. Toxic effects of brominated flame retardants in man and in wildlife. *Environ. Int.* **2003**, *29*, 841. [CrossRef]

56. Botaro, D.; Torres, J.P.M. Difenil éteres polibromados (PBDEs) – Novos poluentes, antigos desafios. *Oecologia Bras.* **2007**, *11*, 167. [CrossRef]

57. Darnerud, P.O.; Lignell, S.; Aune, M.; Isaksson, M.; Cantillana, T.; Redeby, J.; Glynn, A. Time trends of polybrominated diphenylether (PBDE) congeners in serum of Swedish mothers and comparisons to breast milk data. *Environ. Res.* **2015**, *138*, 352. [CrossRef]

58. Ikonomou, M.G.; Addison, R.F. Polybrominated diphenyl ethers (PBDEs) in seal populations from eastern and western Canada: An assessment of the processes and factors controlling PBDE distribution in seals. *Mar. Environ. Res.* **2008**, *66*, 225. [CrossRef]

59. Pestana, C.R.; Borges, K.B.; Da Fonseca, P.; De Oliveira, D.P. Risco ambiental da aplicação de éteres de difenilas polibromadas como retardantes de chama. *Rev. Bras. Toxicol.* **2008**, *21*, 41.

60. Reistad, T.; Mariussen, E. A Commercial Mixture of the Brominated Flame Retardant Pentabrominated Diphenyl Ether (DE-71) Induces Respiratory Burst in Human Neutrophil Granulocytes In Vitro. *Toxicol. Sci.* **2005**, *87*, 57. [CrossRef]

61. Dorneles, P.R.; Lailson-Brito, J.; Dirtu, A.C.; Weijs, L.; Azevedo, A.F.; Torres, J.P.M.; Malm, O.; Neels, H.; Blust, R.; Das, K.; et al. Anthropogenic and naturally-produced organobrominated compounds in marine mammals from Brazil. *Environ. Int.* **2010**, *36*, 60. [CrossRef]

62. Quinete, N.; Lavandier, R.; Dias, P.; Taniguchi, S.; Montone, R.; Moreira, I. Specific profiles of polybrominated diphenylethers (PBDEs) and polychlorinated biphenyls (PCBs) in fish and tucuxi dolphins from the estuary of Paraíba do Sul River, Southeastern Brazil. *Mar. Pollut. Bull.* **2011**, *62*, 440. [CrossRef]

63. Hites, R.A. Polybrominated Diphenyl Ethers in the Environment and in People: A Meta-Analysis of Concentrations. *Environ. Sci. Technol.* **2004**, *38*, 945. [CrossRef]

64. Schecter, A.; Harris, T.R.; Shah, N.; Musumba, A.; Päpke, O. Brominated flame retardants in US food. *Mol. Nutr. Food Res.* **2008**, *52*, 266. [CrossRef]

65. Prefeitura Ribeirão Preto. Prefeitura faz limpeza geral na Lagoa do Saibro. Available online: http://www.maisribeiraopreto.com.br/noticias/prefeitura-faz-limpeza-geral-na-lagoa-do-saibro-2345 (accessed on 11 May 2018).

66. Melymuk, L.; Robson, M.; Csiszar, S.A.; Helm, P.A.; Kaltenecker, G.; Backus, S.; Bradley, L.; Gilbert, B.; Blanchard, P.; Jantunen, L.; et al. From the City to the Lake: Loadings of PCBs, PBDEs, PAHs and PCMs from Toronto to Lake Ontario. *Environ. Sci. Technol.* **2014**, *48*, 3732. [CrossRef]

67. Xiang, N.; Zhao, X.; Meng, X.Z.; Chen, L. Polybrominated diphenyl ethers (PBDEs) in a conventional wastewater treatment plant (WWTP) from Shanghai, the Yangtze River Delta: Implication for input source and mass loading. *Sci. Total Environ.* **2013**, *461–462*, 391–396. [CrossRef]

68. Cincinelli, A.; Martellini, T.; Misuri, L.; Lanciotti, E.; Sweetman, A.; Laschi, S.; Palchetti, I. PBDEs in Italian sewage sludge and environmental risk of using sewage sludge for land application. *Environ. Pollut.* **2012**, *161*, 229. [CrossRef]

69. Ok, G.; Shirapova, G.; Matafonova, G.; Batoev, V.; hyung Lee, S. Characteristics of PAHs, PCDD/Fs, PCBs and PBDEs in the sediment of Lake Baikal, Russia. *Polycycl. Aromat. Compd.* **2013**, *33*, 173. [CrossRef]

70. Mariani, G.; Canuti, E.; Castro-Jiménez, J.; Christoph, E.H.; Eisenreich, S.J.; Hanke, G.; Skejo, H.; Umlauf, G. Atmospheric input of POPs into Lake Maggiore (Northern Italy): PBDE concentrations and profile in air, precipitation, settling material and sediments. *Chemosphere* **2008**, *73*, S114. [CrossRef]

71. Lavandier, R.; Quinete, N.; Hauser-Davis, R.A.; Dias, P.S.; Taniguchi, S.; Montone, R.; Moreira, I. Polychlorinated biphenyls (PCBs) and Polybrominated Diphenyl ethers (PBDEs) in three fish species from an estuary in the southeastern coast of Brazil. *Chemosphere* **2013**, *90*, 2435. [CrossRef]

72. Pozo, K.; Kukučka, P.; Vaňková, L.; Přibylová, P.; Klánová, J.; Rudolph, A.; Banguera, Y.; Monsalves, J.; Contreras, S.; Barra, R.; et al. Polybrominated Diphenyl Ethers (PBDEs) in Concepción Bay, central Chile after the 2010 Tsunami. *Mar. Pollut. Bull.* **2015**, *95*, 480. [CrossRef]

73. Hu, G.; Xu, Z.; Dai, J.; Mai, B.; Cao, H.; Wang, J.; Shi, Z.; Xu, M. Distribution of polybrominated diphenyl ethers and decabromodiphenylethane in surface sediments from Fuhe River and Baiyangdian Lake, North China. *J. Environ. Sci.* **2010**, *22*, 1833. [CrossRef]

74. Wang, X.-T.; Chen, L.; Wang, X.-K.; Zhang, Y.; Zhou, J.; Xu, S.-Y.; Sun, Y.-F.; Wu, M.-H. Occurrence, profiles, and ecological risks of polybrominated diphenyl ethers (PBDEs) in river sediments of Shanghai, China. *Chemosphere* **2015**, *133*, 22. [CrossRef]

75. Samara, F.; Tsai, C.W.; Aga, D.S. Determination of potential sources of PCBs and PBDEs in sediments of the Niagara River. *Environ. Pollut.* **2006**, *139*, 489. [CrossRef]

76. Bradley, P.W.; Wan, Y.; Jones, P.D.; Wiseman, S.; Chang, H.; Lam, M.H.W.; Long, D.T.; Giesy, J.P. PBDEs and methoxylated analogues in sediment cores from two Michigan, USA, inland lakes. *Environ. Toxicol. Chem.* **2011**, *30*, 1236. [CrossRef]

77. Ikonomou, M.G.; Rayne, S.; Addison, R.F. Exponential Increases of the Brominated Flame Retardants, Polybrominated Diphenyl Ethers, in the Canadian Arctic from 1981 to 2000. *Environ. Sci. Technol.* **2002**, *36*, 1886. [CrossRef]

water

MDPI

Article

Removal of Platinum and Palladium from Wastewater by Means of Biosorption on Fungi *Aspergillus* sp. and Yeast *Saccharomyces* sp.

Beata Godlewska-Żyłkiewicz *[iD], **Sylwia Sawicka** and **Joanna Karpińska**[iD]

Institute of Chemistry, University of Bialystok, K. Ciołkowskiego 1K, 15-245 Białystok, Poland
* Correspondence: bgodlew@uwb.edu.pl

Received: 5 June 2019; Accepted: 14 July 2019; Published: 23 July 2019

Abstract: The emission of platinum group metals from different sources has caused elevated concentrations of platinum and palladium in samples of airborne particulate matter, soil, surface waters and sewage sludge. The ability of biomass of *Aspergillus* sp. and yeast *Saccharomyces* sp. for removal of Pt(IV) and Pd(II) from environmental samples was studied in this work. The pH of the solution, the mass of biosorbent, and contact time were optimized. The Langmuir and Freundlich adsorption isotherms and kinetic results were used for interpretation of the process equilibrium of Pt(IV) and Pd(II) on both microorganisms. The maximal efficiency of retention of Pt(IV) on yeast and fungi was obtained at acidic solutions (pH 2.0 for Pt(IV) and pH 2.5–3.5 for Pd(II)). The equilibrium of the biosorption process was attained within 45 min. The best interpretation for the experimental data was given by the Langmuir isotherm. Kinetics of the Pt and Pd adsorption process suit well the pseudo-second-order kinetics model. Fungi *Aspergillus* sp. shows higher adsorption capacity for both metals than yeast *Saccharomyces* sp. The maximum adsorption capacity of fungi was 5.49 mg g^{-1} for Pt(IV) and 4.28 mg g^{-1} for Pd(II). The fungi possess the ability for efficient removal of studied ions from different wastewater samples (sewage and road run-off water). It was also demonstrated, that quantitative recovery of Pd from industrial wastes could be obtained by biosorption using *Aspergillus* sp.

Keywords: biosorption; precious metals; selective sorbent; isotherm adsorption models; environmental samples; run-off water

1. Introduction

Platinum and palladium due to their corrosion resistance, alloying ability and unique catalytic properties are used in various chemical productions and metallurgy. The electrical conductivity and durability of these metals are exploited in electronic applications for the production of multi-layer ceramic (chip) capacitors, and plating connectors and lead frames. Components inside computers and mobile phones are linked by connectors plated with a conductive layer of precious metal [1]. The production of PC computers, mobile phones and entertainment devices generates electronic waste (e-waste) [2]. However, the most important application of Pt and Pd is connected with the production of three-way catalytic converters for car engines [3]. The emission of these metals into the environment is connected with the operation of vehicle catalysts and their recycling. Elevated concentrations of Pt and Pd have been found in samples of airborne particulate matter, road and tunnel dust, and soil close to the roads [4–6]. Discharges of anthropogenic contamination to surface waters include both atmospheric deposition and stormwater runoff. Analysis of sewage sludge, surface waters, and ditch sediments demonstrated that the anthropogenic activity has resulted in elevated concentrations of these elements in such samples [6–8]. Globally, Nuss and Blengini [9] have found that anthropogenic fluxes of Pd and Pt induced by the EU-28 countries might be greater than the respective global natural fluxes. Therefore, methods and processes for the removal of precious metals from waste have become necessary for the creation of the sustainable world [2]. To decrease the pollution with heavy metals, many processes like adsorption, precipitation, coagulation, ion-exchange, electro-dialysis have been developed [10]. One of the modern environmentally friendly technologies for the recovery of platinum group metals is biosorption [11].

A literature survey shows that microorganisms and native biomaterials, such as industrial and agricultural wastes and compounds derived from plant and animal tissues (e.g., lignin, tannin, chitin and chitosan), have been used as effective metal sorbents [11–15]. The majority of studies devoted to biosorption of Pt and Pd have been performed on biomass derived from plants and marine organisms [16–22] with various derivatives of chitosan preferred [16–18]. Very few microorganisms (some strains of bacteria, algae, fungi) [23–30] and *Tobacco mosaic* virus [13] have been used for this purpose. The recovery of platinum by a poly(ethylenimine) (PEI) modified biomass, prepared by attaching PEI onto the surface of inactive *Escherichia coli* biomass was reported by Won et al. [23]. This sorbent was tested for the removal of Pt from wastewater collected from an industrial laboratory. In the next paper Won et al. [24] used PEI-modified *Corynebacterium glutamicum* for recovery of Pd from the hydrochloric acid solution. Different species of *Desulfovibrio*, sulfate-reducing bacteria, have an ability to remove Pd(II) at pH 3 with its further reduction to Pd(0). The potential of such bacteria was also shown for the recovery of Pt and Pd from spent automotive catalyst leachates using hydrogen as the reductant [25]. Turner et al. [28] studied the uptake of platinum group elements by marine macroalgae, *Ulva lactuta* in sea water. Algae *Chlorella vulgaris* immobilized on cellulose have been used for the selective separation of Pt and Pd from environmental matrices [29]. A sulfothermophilic red microalga, *Galdieria sulphuraria*, living in hot sulfur springs, was used for simultaneous removal of gold and palladium from model wastewater [30]. Competitive biosorption of Pt(IV) and Pd(II) by *Escherichia coli* [26] and *Providencia vermicola* [27] was studied in model solutions.

Saccharomyces sp. is a single-cell yeast being used for the biofuel, bakery and beverage industries or for the production of biotechnological products. Such biomass possesses an ability to accumulate a broad range of heavy metals under a wide range of external conditions [31]. Mack et al. [32] have studied a kinetic of the sorption of Pt by immobilized *Saccharomyces cerevisiae* under acidic conditions. Godlewska-Żyłkiewicz [33] observed the highest retention of platinum (62%–65%) and palladium (95%) on free cells of baker's yeast at pH range from 1.6 to 2.2. A solid-phase extraction procedure using yeast *S. cerevisiae* immobilized in calcium alginate beads was proposed for the determination of Pd in road dust by electrothermal atomic absorption spectrometry (ETAAS) [34]. Yeast *S. cerevisiae* immobilized on a cellulosic resin was also used for selective on-line separation of Pt(IV) from river water prior to its chemiluminescent determination [35].

The fungal microorganisms are used extensively in a variety of large-scale industrial fermentation processes and production of gluconic acid, citric acid and many enzymes [36]. Waste biomass of free and immobilized cells of *Aspergillus* was used to remove heavy metal ions, such as cadmium, lead, chromium, iron and nickel from different matrices [37–39]. The first report on the bioaccumulation of platinum and other metals by fungi was published by Moore et al. [40]. The uptake efficiency of Pt equal to 85% at pH 2–3 within 48 h was reported. It was also demonstrated that *Aspergillus* sp. immobilized on cellulose resin Cellex-T could be used as a selective sorbent for solid-phase extraction of Pt and Pd from the complex matrix of road dust [41].

In this work, the ability of biomass of *Aspergillus* sp. and yeast *Saccharomyces* sp. for biosorption of Pt(IV) and Pd(II) from chloride aqueous solutions was studied. Studies comparing different sorbents will reveal the importance of metal-biomass specificity. The pH of the solution, the mass of biosorbent, and contact time have been studied in order to find the optimal parameters for biosorption of metals. The Langmuir and Freundlich adsorption isotherms have been used for interpretation of the process equilibrium of Pt(IV) and Pd(II) on *Aspergillus* sp. and *Saccharomyces* sp. The sorption kinetics of both ions on these microorganisms was also studied. The method was tested for the removal of these precious metals from different wastewater samples (sewage and road run-off water). Moreover, it was demonstrated that the quantitative recovery of Pd from industrial wastes (anode slime) might be obtained using fungal biosorbent.

2. Materials and Methods

2.1. Instrumentation and Measurement Conditions

A Solaar M6 (Thermo Electron Corporation, Gloucester, UK) atomic absorption spectrometer equipped with an electrothermal atomizer (ETAAS) and a Zeeman background correction system was used. Hollow cathode lamps (CPI International, Santa Rosa, USA) were operated at a current of 10 mA for Pt and 6 mA for Pd. The absorbance signals were measured with 0.5 nm spectral bandpass at 265.9 nm and 247.6 nm for Pt and Pd, respectively. Pyrolytically coated graphite tubes were used for atomization of analytes. The time/temperature program for the Pd determination was: drying at 100 °C for 30 s, 110 °C for 20 s, ashing at 1100 °C for 20 s and atomization at 2200 °C for 3 s. The time/temperature program for the Pt determination was: drying at 100 °C for 30 s, 110 °C for 20 s, ashing at 1200 °C for 20 s and atomization at 2500 °C for 3 s.

An inoLab pH Level 1 (WTW, Weilheim, Germany) pH meter equipped with an electrode SenTix 21 (WTW, Weilheim, Germany) was used to measure the pH. A centrifuge MPW 312 (MPW Med. Instruments, Warszawa, Poland) was used for separation of biomass from the supernatant. A Hitachi Model S-3000N (Hitachi, Tokyo, Japan) scanning electron microscope equipped with an energy dispersive X-ray microanalysis (EDX) was used to detect metals in the cells.

2.2. Reagents and Solutions

Standard solutions of Pt as a hexachloroplatinic (IV) acid (30%) (POCH, Gliwice, Poland) and Pd as $PdCl_2$ (1 g L^{-1}) (SPC SCIENCE, Baie D'Urfé, Quebec, Canada) were used. Stock solutions of Pd(II) and Pt(IV) (1 μg L^{-1}) in 1 mol L^{-1} HCl were prepared daily from standard solutions. The appropriate working solutions of Pd(II) and Pt(IV) used for biosorption studies were prepared by dilution with Milli-Q water (Millipore, Burlington, USA). Hydrochloric acid (37% Trace Select, Fluka, Lyon, France) and sodium hydroxide (analytical grade, Standard, Lublin, Poland) were used for adjustment of pH of solutions. Nitric acid (65% Trace Select, Fluka, Lyon, France) and hydrochloric acid (37% Trace Select, Fluka, Lyon, France) were used for digestion of samples.

A growth media were Czapek dox agar (Fluka, Buchs, Switzerland) containing sucrose (30 g L^{-1}), agar (15 g L^{-1}), $NaNO_3$ (3 g L^{-1}), K_2HPO_4 (1 g L^{-1}), KCl (0.5 g L^{-1}), $MgSO_4 \cdot 7H_2O$ (0.5 g L^{-1}), $FeSO_4 \cdot 7H_2O$ (0.01 g L^{-1}), Yeast extract (Fluka, Buchs, Switzerland) (a mixture of amino acids, peptides, water soluble vitamins and carbohydrates), YPG agar (Fluka, Buchs, Switzerland) and α-D-glucose (Aldrich, Hamburg, Germany).

2.3. Preparation of Cells

Aspergillus sp. isolated from the soil was prepared according to the procedure described in Reference [42]. In short, it was inoculated on solid nutrient medium onto Petri dishes and aerobically incubated at 30 °C for 72 h. The growth medium: 50 g L^{-1} of Czapek 146 dox agar and 3.89 g L^{-1} of Yeast extract was sterilized by autoclaving (20 min at 121 °C). *Saccharomyces* sp. (*Saccharomyces cerevisiae*, Baker's yeast, type II, (Sigma-Aldrich, Hamburg, Germany) was grown aerobically at 37 °C for 96 h in a pre-sterilized (20 min at 121 °C) solid medium containing: yeast extract, 1% (w/v); peptone, 2% (w/v); glucose, 2% (w/v) and agar, 2% (w/v). Biomass of fungi and yeast was scraped from the growth medium and then washed with 5 mL of 0.12 mol L^{-1} HCl and 5 mL of Milli-Q water in order remove growth solution residues and to stabilize the surface activity.

2.4. Biosorption Studies

The biosorption of Pt(IV) and Pd(II) was studied under batch experimental conditions using 0.1 g of wet mass (WM) of biosorbent (fungal or yeast cells) and 5 mL of 0.1 mg L^{-1} of Pt(IV) or

0.075 mg L^{-1} Pd(II) solutions. The pH of Pt(IV) or Pd(II) solutions was adjusted to the desired value with 0.1 mol L^{-1} HCl or 0.1 mol L^{-1} NaOH. Next, 0.1 g of wet biomass was added to 5 mL of such solution. The suspension of biomass was stirred for 60 min at temperature (25 ± 1) °C on a magnetic stirrer and next centrifuged at 4000 rpm (550× g) for 10 min. Concentrations of Pd(II) and Pt(IV) in supernatants were determined by ETAAS using external calibration graph. In some cases, the pH of supernatants was also measured. The effect of initial pH on the retention of analyte on biomass was studied in the pH range from 0.5 to 11. The optimum pH of the Pt solutions was equal to 2 (for fungi and yeast), while the optimum pH of the Pd solutions was 2.5 for yeast or 3.5 for fungi. These pH values were used throughout all biosorption experiments. The effect of biosorbent dosage was studied in the range from 0.025 to 0.5 g (wet mass). The contact time was varied between 5 min and 24 h.

To examine the adsorption model, the solutions of Pt and Pd of initial concentration between 0.01 mg L^{-1} and 20 mg L^{-1} and optimal pH (pH 2 for Pt(IV) and pH 2.5 and 3.5 for Pd(II) on yeast and fungi, respectively) were stirred with 0.1 g of wet biomass for 45 min at 25 °C. Afterward, the samples were centrifuged at 4000 rpm for 10 min. The initial and final concentrations of Pt and Pd in solutions were determined by ETAAS. Each experiment was conducted in triplicate. In order to recalculate the obtained results for dry biosorbent mass, five portions of wet biomass (0.4 g) were dried at 60 °C overnight [37,38] and weighted. It was calculated that 0.1 g of wet biomass is equal to 0.013 g of dry biomass.

The efficiency of retention of analyte on biomass and metal uptake by biomass were calculated from the following equations [43]:

$$E(\%) = \frac{C_0 - C_e}{C_0} \times 100\%. \tag{1}$$

$$q_{exp} = \frac{V(C_0 - C_e)}{w}. \tag{2}$$

where E is the efficiency of analyte retention (in %), C_0 is the initial concentration of metal in solution (mg L^{-1}), C_e is the equilibrium concentration of metal in solution after the biosorption process (mg L^{-1}), q_{exp} is the experimental equilibrium uptake (mg g^{-1} of dry weight), V is the solution volume, and w is the dry weight of biosorbent (g).

Modeling of the isotherm data was attempted using the Langmuir [44] and Freundlich [45] models, which are represented by the following equations:

$$\text{Langmuir model}: q_e = \frac{q_{max}\, bC_e}{1 + bC_e} \tag{3}$$

$$\text{Freundlich model}: q_e = K_F C_e^{\frac{1}{n}} \tag{4}$$

where q_e is the amount of adsorbed metal (mg g^{-1}), C_e is the equilibrium (final) concentration of the metal in solution (mg L^{-1}), q_{max} is the maximum monolayer sorption capacity (mg g^{-1}), b is the Langmuir equilibrium constant (L mg^{-1}), K_F is an empirical constant that provides an indication of the adsorption capacity of biomass, and n is the Freundlich constant, that indicates the intensity of adsorption.

The sorption isotherms were plotted by varying the analyte uptake (q) by biosorbents to the final concentration of Pt or Pd in solution after the biosorption process (C$_e$).

Kinetic models are used to identify the adsorption mechanism type in a studied system. Moreover, kinetics studies are necessary to define the optimum conditions for the metal removal process. The pseudo-first and pseudo-second-order models are most often used for studies of biosorption kinetics of platinum group metals [27].

Linear pseudo-first-order model used in this study was:

$$\ln(q_e - q_t) = \ln q_e - k_1 t \tag{5}$$

Linear pseudo-second-order model was:

$$\frac{t}{q_t} = \frac{1}{k_2\, q_e^2} + \frac{t}{q_e} \tag{6}$$

where q_t and q_e (mg g^{-1}) are the amounts of adsorbed ion a given time and at equilibrium state, respectively, k_1—is the first-order kinetic rate, k_2—is the second-order kinetic rate (g mg^{-1} min^{-1}).

2.5. SEM-EDX Analysis

Scanning electron microscopy with energy dispersive X-ray (SEM-EDX) provides information about the presence of various elements in the biosorbent. In this work, 5 mL of Pt(IV) or Pd(II) solutions of 20 mg L^{-1} (100 μg) were added separately to 0.1 g (wet mass) of fungal cells and gently stirred for 72 h. Next, the cells were centrifuged and the supernatant was discarded. Before SEM-EDX analysis all samples were frozen in liquid nitrogen.

2.6. Samples Used for Recovery Studies

Samples of sewage were obtained from the sewage treatment plant in Bialystok (Poland). Samples of road run-off were taken from the retention reservoir at the ring road of Bialystok. Pt (100 ng mL^{-1}) and Pd (75 ng mL^{-1}) was added to the samples and left for 2 h for equilibration. Then samples were filtered through PVDF discs (0.45 μm, Whatman, Maidstone, Wielka Brytania), adjusted to the required pH with 0.1 mol L^{-1} HCl and stirred with 0.1 g of wet biomass for 2 h. The biomass was separated by centrifugation (10 min, 4000 rpm), washed with 0.01 mol L^{-1} HNO$_3$ and digested in quartz crucibles with concentrated HNO$_3$ on a laboratory heater. The digested samples were diluted appropriately with MQ water before determination of metals by ETAAS.

The samples of anode slime (200 mg) obtained from the Institute of Non-Ferrous Metals in Gliwice (Poland) were digested in Teflon vessels in a closed digestion microwave system ETHOS PLUS (Milestone, Sorisole, Italy) with 8 mL of *aqua regia*. The residue was separated and solutions were transferred into quartz crucibles and evaporated at a hot-place near to dryness with concentrated HCl (3 × 2 mL) and next diluted with Milli-Q water to 15 mL. Samples were filtered through PVDF filters, adjusted to the required pH and stirred with 0.1 g of wet biomass for 2 h. Next, the biomass was separated by centrifugation (10 min 4000 rpm) and concentrations of Pt and Pd in supernatants were determined by ETAAS.

3. Results and Discussion

3.1. Effect of pH on Biosorption

The effect of an initial pH on the biosorption of Pt(IV) and Pd(II) on fungal and yeast biomass was studied at the pH range from 0.5 to 11.0 (Figure 1). The highest efficiency of biosorption of Pt(IV) ions on yeast cells was observed at the pH range from 1.8 to 3.5, while on fungal cells was nearly quantitative (90%–96%) at the pH range from 2.0 to 11.0. The maximal efficiency of biosorption of Pd(II) ions on yeast (85%) was obtained at a very narrow range of pH from 2.0 to 3.0. Low biosorption of analyte occurred both in a strong acidic medium and in solutions of pH above 4.0. The biosorption of Pd(II) on fungal cells reached the highest values at a pH range of 4.0–11.0. In all cases, the biosorption of Pt(IV) and Pd(II) was lower in strong acidic solutions (at pH < 1.8). It is apparent that different attractions are responsible for biosorption of metals on the studied microorganisms. Further experiments on biosorption of Pt(IV) ions were performed at pH 2 on yeast and fungal cells, while of Pd(II) ions at pH 2.5 on yeast and pH 3.5 on fungal cells.

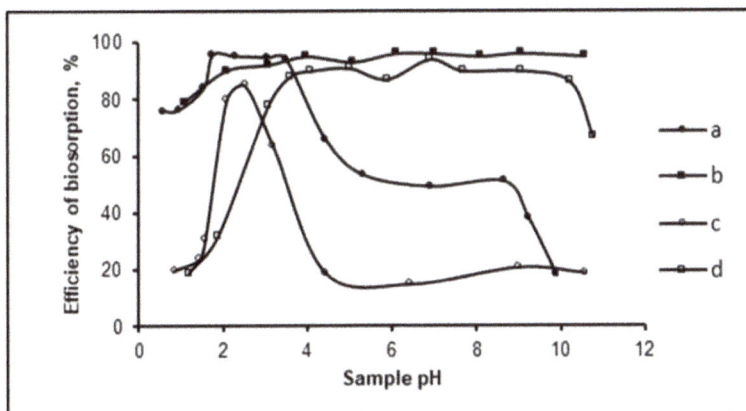

Figure 1. Influence of sample pH on the efficiency of biosorption of Pt (0.1 mg L^{-1}) and Pd (0.075 mg L^{-1}) on yeast *Saccharomyces* sp. and fungi *Aspergillus* sp. (contact time 60 min): a—Pt(IV) on yeast; b—Pt(IV) on fungi; c—Pd(II) on yeast; d—Pd(II) on fungi.

Several researchers have also investigated the effect of pH on the biosorption of Pt(IV) and Pd(II) using different microbial mass and comparable results have been reported (Table 1). For instance, biosorption of Pt(IV) [32,33] and Pd(II) [33] on free cells of baker's yeast was maximal in acidic media. The similar effect of pH on the Pd(II) biosorption was observed on yeast immobilized in calcium alginate. The biosorption of Pd(II) ions on this sorbent reached the highest values at a pH range 1.0–2.5 and significantly decreased for less acidic solutions [34].

During the course of the experiments, the pH of the suspensions before and after the biosorption process of each metal was also measured. It was found that the pH of Pt(IV) and Pd(II) solutions incubated with both microorganisms at pH 2–2.5 slightly increased (ΔpH ~0.14), while the pH of suspension incubated without metal ions maintained almost constant (ΔpH $\leq$ 0.05). A dissimilar effect was observed during biosorption of Pd(II) ions by fungal cells (at pH 3.5), as the pH of suspension decreased by 0.4 pH unit, while the pH of suspension incubated without metal ions decreases only by 0.3 pH unit. These phenomena should be discussed in terms of reactions of Pt(IV) and Pd(II) ions in aqueous solutions.

It is known that the chemical form of platinum group metals in solutions at the equilibrium state depends on the medium type, the concentration of chloride ions and temperature. In strong acidic media anionic chlorocomplexes of Pt(IV) (as $PtCl_6^{2-}$) and Pd(II) (as $PdCl_4^{2-}$) predominate. With the decrease in solution's acidity, the proceeding aquation and hydrolysis reactions cause a formation of different aquachloro- and aquahydroxocomplexes of platinum and palladium ($PtCl_5(H_2O)^-$, $PdCl_4^{2-}$, $PdCl_3(H_2O)^-$, $PdCl^+$ and $Pd(OH)_2$), as was shown in References [19,46]. Additionally, the chemical composition and acidity of the solution influence the activity and accessibility of functional groups present at the surface of the cell's wall.

Table 1. Biosorption conditions and parameters of isotherm adsorption models of Pd(II) and Pt(IV) on various microorganisms.

Microorganism	Biosorption Conditions	Parameters of Equilibrium Isotherm Models		Application	Ref.
		Langmuir	Freundlich		
Tobacco mosaic virus	Sample pH: 5–5.5 contact time: 1 h biomass: 0.038 g L^{-1} temperature: 50 °C	(Pd) R^2: 0.95 q_{max}: 368.21 mg g^{-1} b: 0.36 L mg^{-1}	-	Standard solutions	[13]
Tobacco mosaic virus-wild		(Pd) R^2: 0.98 q_{max}: 312.87 mg g^{-1} b: 0.12 L mg^{-1}			
Escherichia coli	Sample pH: 3 contact time: 24 h biomass: 0.09 g temperature: 25 °C	(Pd) R^2: 0.991 q_{max}: 141.1 mg g^{-1} b: 0.014 L mg^{-1}	-	Standard solutions	[14]
Polyallylamine hydrochloride modified Escherichia coli		(Pd) R^2: 0.969 q_{max}: 265.3 mg g^{-1} b: 0.042 L mg^{-1}	-		
Polyethylenimine (PEI)-modified Escherichia coli	Sample pH: extremely acidic condition contact time: 60 min biomass: 0–1.8 g temperature: 25 °C	(Pt) R^2: 0.965 q_{max}: 108.8 mg g^{-1} b: 0.0014 L mg^{-1}	(Pt) R^2: 0.952 K$_F$: 1.465 L g^{-1} n: 1.835	ICP wastewater	[23]
Escherichia coli	Sample pH: 1.2 contact time: 24 h biomass: 0.06 g temperature: 25 °C	(Pd) R^2: 0.962 q_{max}: 38.87 mg g^{-1} b: 0.48 L mg^{-1} (Pt) R^2: 0.991 q_{max}: 45.65 mg g^{-1} b: 0.58 L mg^{-1}		Single and binary standard solutions	[26]
Desulfovibrio desulfuricans	Sample pH: 3 contact time: 20 min biomass: 0.0015 g temperature: 30 °C	(Pd) q_{max}: 125.0 mg g^{-1} b: 1.21 L mg^{-1} (Pt) q_{max}: 62.5 mg g^{-1} b: 0.50 L mg^{-1}	(Pd) K$_F$: 69.7 L g^{-1} n: 4.24 (Pt) K$_F$: 24.2 L g^{-1} n: 3.43	Standard solutions	[25]

Table 1. *Cont.*

Microorganism	Biosorption Conditions	Parameters of Equilibrium Isotherm Models		Application	Ref.
		Langmuir	Freundlich		
Desulfovibrio fructodivorans		(Pd) q_{max}: 119.8 mg g^{-1} b: 0.12 L mg^{-1} (Pt) R^2: q_{max}: 32.3 mg g^{-1} b: 1.17 L mg^{-1}	(Pd) K_F: 10.4 L g^{-1} n: 2.35 (Pt) K_F: 20.3 L g^{-1} n: 7.12		[25]
Desulfovibrio vulgaris		(Pd) q_{max}: 106.3 mg g^{-1} b: 0.66 L mg^{-1} (Pt) q_{max}: 32.1 mg g^{-1} b: 0.42 L mg^{-1}	(Pd) K_F: 41.9 L g^{-1} n: 3.73 (Pt) K_F: 16.8 L g^{-1} n: 5.92		[25]
Providencia vermicola	Sample pH: 4 contact time: 3 h biomass: 0.075 g temperature: 30 °C	(Pd) R^2: 0.95 q_{max}: 119 mg g^{-1} b: 0.1 L mg^{-1} (Pt) R^2: 0.97 q_{max}: 30.2 mg g^{-1} b: 0.13 L mg^{-1}	(Pd) R^2: 0.78 K_F: 1.27 L g^{-1} 1/n: 1.07 (Pt) R^2: 0.72 K_F: 0.16 L g^{-1} 1/n: 1.06	Single and binary standard solutions	[27]
Saccharomyces sp.	Sample pH: (Pd) 2.5, (Pt) 2 contact time: 45 min biomass: 0.1 g (wet), 0.013 g (dry) temperature: 25 °C	(Pd) R^2: 0.9874 q_{max}: 0.042 mg g^{-1} b: 0.0204 L mg^{-1} (Pt) R^2: 0.9860 q_{max}: 0.185 mg g^{-1} b: 0.0068 L mg^{-1}	(Pd) R^2: 0.797 K_F: 0.095 L g^{-1} n:1.630 (Pt) R^2: 0.975 K_F: 0.286L g^{-1} n: 1.505	Standard solutions	This study
Aspergillus sp.	Sample pH: (Pd) 3.5, (Pt) 2 contact time: 45 min biomass: 0.1 g (wet), 0.013 g (dry) temperature: 25 °C	(Pd) R^2: 0.9823 q_{max}: 4.277 mg g^{-1} b: 0.0021 L mg^{-1} (Pt) R^2 0.0010 q_{max}: 5.488 mg g^{-1} b: 0.0006 L mg^{-1}	(Pd) R^2: 0.805 K_F: 2.842 L g^{-1} n: 1.485 (Pt) R^2: 0.962 K_F: 1.766 L g^{-1} n: 1.164	Run-off water, sewage, anode slime	

The high biosorption of platinum and palladium on yeast cells at pH range from 1.8 to 3.0 occurs probably through electrostatic attractions between the protonated functional groups of the sorbent and their anionic chlorocomplexes. It was reported that the zeta potential of *Saccharomyces* sp. immobilized on cone biomass was positive at pH 2.0 and the overall surface of the biomass was negatively charged at the pH values between 3.0 and 7.0 [47]. Other studies [48] have shown, that the surface of free cells of yeast was negatively charged for pH higher than 3.5 favoring adsorption of cationic species. Lower sorption of Pt(IV) and Pd(II) observed in strong acidic medium is an effect of competing of chloride ions and anionic chlorocomplexes of platinum and palladium present in the solution for the protonated functional groups of yeast's cell wall. The results presented above imply that the biosorption mechanism could be based on electrostatic attractions between analytes and microorganisms. In our former studies [35] platinum was efficiently (83%) removed from immobilized yeast with 3 mol L^{-1} NaCl, showing that ion-exchange mechanism could be also involved in the metal binding to biomass. However, in our opinion, other kinds of interactions may also participate in biosorption of Pt, as above 91% of platinum was recovered from the same sorbent with an acidic solution of thiourea [34]. A complex mechanism, likewise, initial non-specific sorption of platinum ions due to electrostatic attractions between protonated sorbent and platinum anions followed by chemical sorption of Pt(IV) by *S. cerevisiae* was suggested by Mack et al. [32]. Kim et al. [26] stated that primary amines present in the biomass are responsible for selective biosorption of Pd(II) and Pt(IV) by *E. coli*. Moreover, the affinity of amines toward the Pd(II) was much higher than for Pt(IV) ions.

The attractions between Pt(IV), Pd(II) and fungi cells are also complex in their nature including electrostatic attractions of anionic complexes to the positively charged amino groups (at a lower pH range) and complexation of neutral or cationic forms of their aqua- and aquahydroxochloro-complexes to functional groups of cell wall from neutral and basic solutions. The effect of modification of a fungal cell wall by acidic solution has to be also considered. The initial studies (data not shown) were performed on the effect of modification of fungal cells (washing with MQ water, washing with 0.12 mol L^{-1} HCl and next MQ water, boiling with 0.5 mol L^{-1} NaOH and next washing with MQ water) on the biosorption of platinum and palladium. The experiments showed that biosorption of both metals from solutions of pH > 4 on the fungi washed with water and modified with NaOH was significantly lower (biosorption efficiency from 5% to 40%) than on the fungi modified with an acid solution (biosorption efficiency > 90%). The efficient retention of Pt and Pd on fungi treated by each procedure occurred from solutions of pH 2–4. These results cannot be easily compared to literature data as the research on biosorption of precious metals on fungi is mainly focused on gold [9]. Gold and other heavy metals, e.g., copper, cadmium and lead, are retained on *Aspergillus* sp. biomass in weakly acidic or neutral pH [2,37–39]. Hence, in our further work, in order to achieve better selectivity of removal of precious metals from samples containing other metals, the studies were carried out at pH 2.0 for Pt(IV) and 3.5 for Pd(II). At pH ≥ 3 biosorption of Pd(II) ions in the form of PdCl$^+$ and Pd(OH)$_2$ species [19] may occur. The increase of the concentration of H$^+$ ions in cells suspension after the biosorption process suggests that Pd(II) ions could be bound by an ion-exchange mechanism.

3.2. Effect of Contact Time and Biomass Dosage

The effect of contact time on the biosorption of Pt(IV) and Pd(II) ions on yeast and fungal biomass was studied in the range from 5 min to 24 h. About 60% of Pd(II) was taken up by yeast within the first 10 min, and after 45 min the amount of biosorbed Pd(II) reached a constant value of 90.1% ± 0.7%. The biosorption of Pt(IV) on yeast was faster, as above 93% of the initial amount of ions was retained within 5 min of contact time. These results are consistent with the results of Mack et al. [32], who observed the rapid platinum removal during the first 5 min. Next, the process, identified as chemical sorption, was much slower. The efficiency of biosorption of Pd(II) and Pt(II) on fungi within 15 min reached 97% and 91%, respectively. Figure 2a reveals that over 80% of biosorption of both ions occurred within 15–20 min and the equilibrium was attained within 45 min. The reproducibility of the biosorption process (n = 6) on fungal biomass was 89.0 ± 1.4% for Pt(IV) and 96.0 ± 3.8% for

Pd(II), while on yeast biomass was 94.1 ± 1.6% for Pt(IV) and 84.3 ± 4.7% for Pd(II). There was no remarkable change in the amount of metal taken up after 24 h of contact time. The time required to reach the biosorption equilibrium of other metals ions on fungi *Aspergillus* sp. [37–39] and yeast *Saccharomyces* sp. [47–49] was longer than that found in this work. The equilibrium time between platinum or palladium and moss biomass was comparable [21] but this time was much longer for other biomaterials (in the range 24–96 h) [17,20,22]. In practice, a sorbent with a faster uptake is better for the removal of metals.

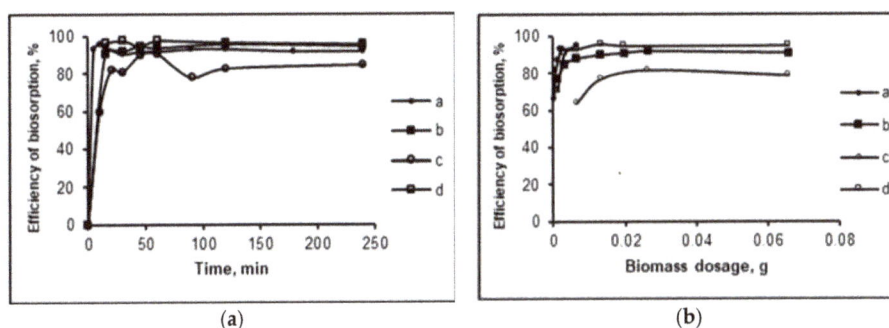

Figure 2. Dependency of the efficiency of biosorption of Pt (0.1 mg L^{-1}) and Pd (0.075 mg L^{-1}) at optimal pH conditions on *Saccharomyces* sp. and *Aspergillus* sp. on (**a**) contact time (**b**) biosorbent mass (contact time 45 min); a—Pt(IV) on yeast; b—Pt(IV) on fungi; c—Pd(II) on yeast; d—Pd(II) on fungi.

The influence of biomass dosage on Pt(IV) and Pd(II) biosorption was studied in the range of 0.001–0.066 g of dry mass (0.025–0.5 g of wet mass) (Figure 2b). The biosorption efficiency of both ions increases along with the increasing mass of biomass to a value of 0.013 g (0.1 g of wet mass). This is probably an effect of a higher number of binding sites on the surface of the biosorbent. The efficiency of biosorption was constant in the range of 0.013–0.066 g of biomass dosage demonstrating the formation of an equilibrium between the ions bound to the biosorbent and those remaining in the solution.

3.3. Biosorption Isotherms

The equilibrium biosorption isotherm is of importance in the design of sorption systems. The Langmuir isotherm is based on the monolayer adsorption on the active sites of the adsorbent. The Freundlich isotherm explains the adsorption on a heterogeneous (multiple layers) surface with uniform energy. Although the empirical models cannot provide any mechanistic understanding of the adsorption phenomena, these models may be used to conveniently estimate the maximum uptake of precious metals from experimental data.

The experimental sorption isotherms obtained under optimal conditions and the sorption isotherms predicted by the Langmuir and Freundlich models along with the determination coefficients (R^2) are shown in Figure 3. The calculated Langmuir and Freundlich parameters are listed in Table 1. The values of R^2, which were found to be above 0.982, indicating that the biosorption of Pt and Pd on the fungi and yeast biomass is consistent with the Langmuir model. In other words, monolayer adsorption took place at the binding sites of both microorganisms. Both types of microorganisms have a higher affinity for Pd(II) than for Pt(IV) ions. The highest affinity of the sorbent for the sorbate was observed between the yeast and the Pd(II) ions (the highest b constant). Fungi *Aspergillus* sp. show a higher adsorption capacity then yeast *Saccharomyces* sp. for both metals. According to the Langmuir model, the fungi achieved a maximum platinum uptake of 5.488 mg g^{-1} that was 30-fold higher than that of the yeast biomass (0.185 mg g^{-1}). The specific palladium sorption capacity of fungi (4.277 mg g^{-1}) was 100-fold higher than that of yeast. This is most likely related to the composition of the cell wall of *Aspergillus* sp., which enhances the adsorption capacity. As can be seen, fungi exhibit an advantage in the removal of Pt and Pd compared with yeast. Hence, further studies were performed on *Aspergillus* sp. cells. The specified sorption of Pt(IV) and Pd(II) to fungi cells was confirmed by EDX analysis (Figure 4b,c). The signal of Pt is seen at 2 keV and Pd at 2.8 keV and 0.2 keV (covered by the peak of C).

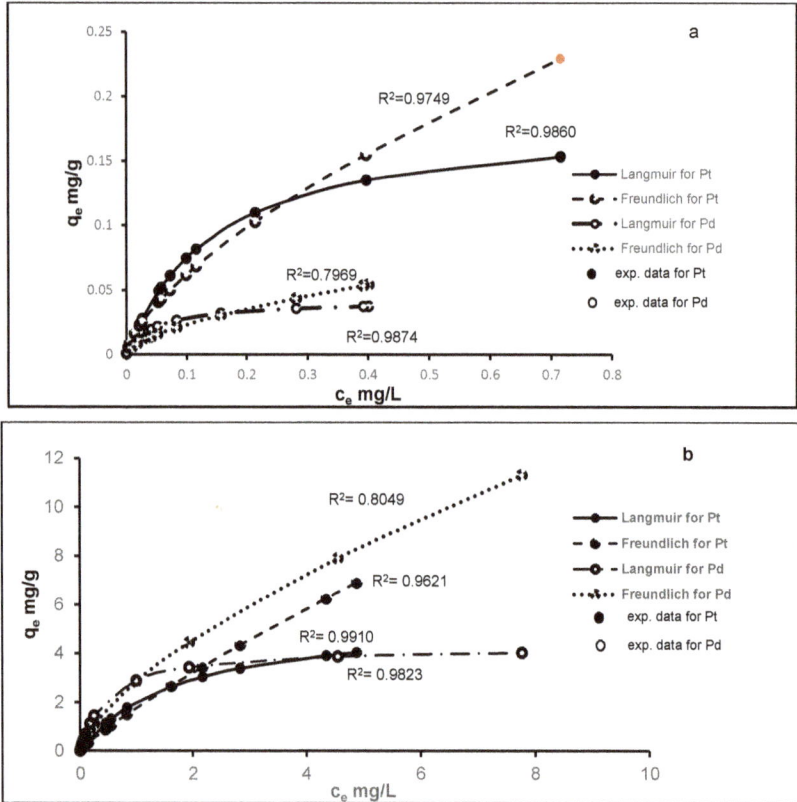

Figure 3. Sorption isotherms of Pt and Pd on (**a**) *Saccharomyces* sp. and (**b**) *Aspergillus* sp. Biosorption of Pt on yeast and fungi at pH 2, biosorption of Pd on yeast at pH 2.5, on fungi at pH 3.5, dry biomass 0.013 g, contact time 45 min.

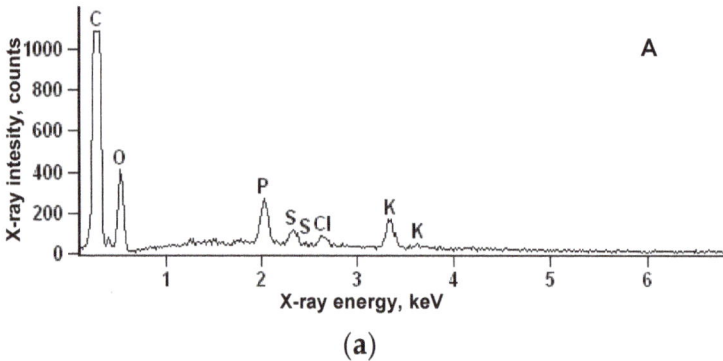

(a)

Figure 4. *Cont.*

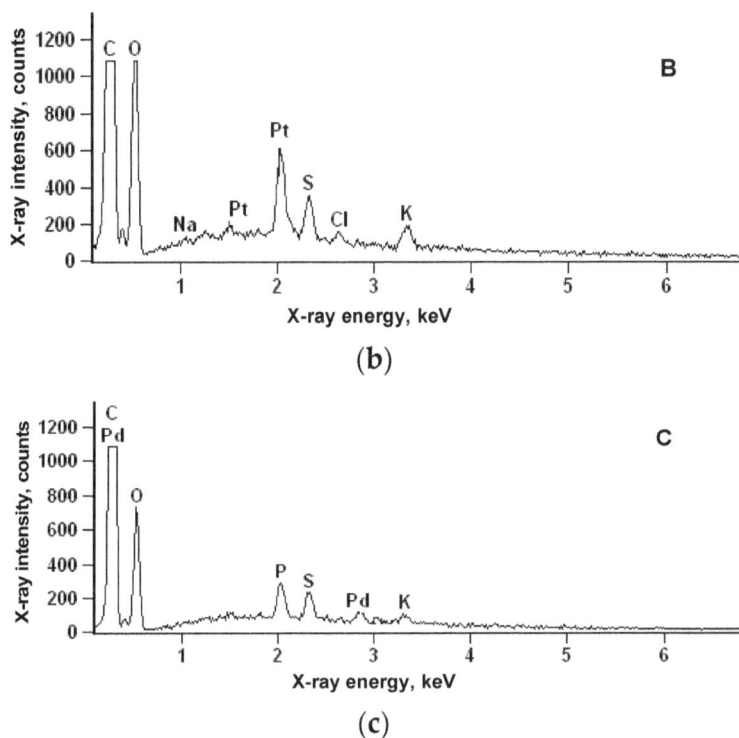

Figure 4. SEM-EDS spectra of (a) free fungi cells, (b) fungi after incubation with Pt(IV), (c) fungi after incubation with Pd(II).

The review of biosorption conditions and parameters of isotherm adsorption models of Pd(II) and Pt(IV) on various microorganisms as well as results obtained in this work are given for comparison in Table 1. As can be seen, the maximum biosorption capacities of fungi observed in our experiment are lower in comparison with the sorption capacities of other microorganisms. However, the time required to obtain the equilibrium of the biosorption process was shorter than in other works.

3.4. Biosorption Kinetics

In order to complete evaluation of the sorption mechanism of examined biomaterials, the kinetic parameters of the sorption process were determined. For this purpose, sorption studies of Pt or Pd ions were carried out under optimal conditions using 0.1 g (wet biomass) of the currently tested biosorbent. The applied concentration of Pt was 0.1 mg L^{-1}, while Pd was 0.075 mg L^{-1}. Every 15 min an appropriate amount of the tested suspension was taken and the current content of the studied ion was determined. The pseudo-first (PFO) and -second (PSO) models of adsorption kinetics were assumed and checked [50]. Basic kinetic parameters such as the amount of adsorbed metal ions at equilibrium (q_e) and the adsorption rate constants (k_1 and k_2) were determined experimentally using the graphical dependence of log ($q_e - q_t$) vs. *t* for the PFO and $1/q_t$ vs. $1/t$ for the PSO models of adsorption kinetics. The suitability of experimental data to the assumed model of adsorption kinetics was determined based on the regression coefficient. The obtained kinetic parameters of examined processes are presented in Table 2.

The obtained data show that the kinetics of the Pt and Pd adsorption process suit well the pseudo-second order kinetics model. This means that rate-limiting step most likely involves chemical interactions leading to the binding of metal ions to the adsorbent surface such as, among others, ion exchange or complexing processes [50]. This conclusion confirms results obtained by the examination of adsorption isotherms. The mass of adsorbed ion at equilibrium state calculated for the dry mass of biosorbent was 7.819×10^{-2} mg g^{-1} and 5.864×10^{-2} mg g^{-1} for Pt and Pd ions, respectively.

Table 2. The kinetic parameters of Pt and Pd adsorption by examined biosorbents (q_e—the amount of adsorbed ion at equilibrium state (mg g^{-1}), k_1—the first order kinetic rate (min^{-1}); k_2—second order kinetic rate (g mg^{-1} min^{-1}).

Studied Ion	Microrganism	Pseudo-First Order Kinetics Model		Pseudo-Second Order Kinetics Model	
Pt	*Saccharomyces* sp.	q_e	4.670×10^{-3}	q_e	4.601×10^{-3}
		k_1	8×10^{-5}	k_2	830.173
		R^2	0.145	R^2	0.999
	Aspergillus sp.	q_e	0.358×10^{-4}	q_e	4.122×10^{-3}
		k_1	2.2×10^{-3}	k_2	210.474
		R^2	0.373	R^2	0.999
Pd	*Saccharomyces* sp.	q_e	1106.0	q_e	3.188×10^{-3}
		k_1	5.1×10^{-3}	k_2	226.254
		R^2	0.515	R^2	0.990
	Aspergillus sp.	q_e	3.218×10^{-3}	q_e	3.092×10^{-3}
		k_1	2×10^{-4}	k_2	280.620
		R^2	0.491	R^2	0.996

3.5. Recovery of Platinum and Palladium from Wastes

The presence of matrix components of the samples, inorganic ions as well as organic matter, may influence the specific biosorption of analyte. Hence, in this work, the recovery of platinum and palladium ions from samples containing complex matrix was studied. As the content of Pt and Pd in samples of crude sewage obtained from municipal sewage treatment plant and road run-off was not determined, they were spiked with different amounts of analytes and left for equilibration. Next, they were pretreated and analyzed according to the procedure described in Section 2.6. The efficiencies of retention of analytes and their recoveries from biomass were calculated and the results are given in Tables 3 and 4.

Table 3. Application of fungi *Aspergillus* sp. for separation of Pt(IV) from various samples (sample pH 2, mass of biosorbent 0.1 g, contact time 2 h, washing agent: 0.01 mol L^{-1} HNO$_3$, n = 3).

Sample Type	Pt(IV) Amount, µg	Retention ± SD, %	Recovery ± SD, %
Standard solution	100	89.1 ± 1.1	70.9 ± 2.5
	200	81.3 ± 0.8	65.3 ± 4.3
	500	68.1 ± 6.0	66.5 ± 4.3
	1000	56.5 ± 1.7	51.3 ± 5.2
Road run-off	0.5	95.6 ± 0.9	68.9 ± 2.1
	100	90.7 ± 0.7	62.9 ± 6.2
	200	82.0 ± 0.9	51.2 ± 0.1
	500	70.7 ± 5.9	50.9 ± 4.7
	1000	47.4 ± 0.0	-
Sewage	0.5	62.3 ± 7.5 [a]	37.5 ± 3.4 [a]
	0.5	89.4 ± 0.5	82.7 ± 3.4
	1000	36.7 ± 0.0	-
Anode slime	1.2 [b]	-	74.8 ± 5.7

[a] mass of biomass 0.013 g, [b] constituent content of anode slime: Pt—6 µg g^{-1}, Ag—40%, Pb—30%, Sb—1%–3%, Se—1%–2%, Cu—1%–2%, S—6%, Au—200 µg g^{-1}, Pd—7 µg g^{-1}.

Table 4. Application of fungi *Aspergillus* sp. for separation of Pd(II) from various samples (sample pH 3.5, mass of biosorbent 0.1 g, contact time 2 h, washing agent: 0.01 mol L^{-1} HNO$_3$, n = 3).

Sample Type	Metal Amount, µg	Retention ± SD, %	Recovery ± SD, %
Standard solution	1000	98.1 ± 0.2	95.2 ± 3.3
Road run-off	0.5	93.7 ± 0.2	98.8 ± 3.4
	1000	99.4 ± 0.0	82.4 ± 4.8
Sewage	0.5	99.4 ± 0.3 [a]	95.8 ± 5.1 [a]
	0.5	91.2 ± 1.0	86.4 ± 1.1
	1000	88.6 ± 0.9	65.7 ± 6.2
Anode slime	1.4 [b]	-	97.9 ± 7.7

[a] mass of biomass 0.013 g, [b] constituent content of anode slime: Pd—7 µg g^{-1}, Ag—40%, Pb—30%, Sb—1%–3%, Se—1%–2%, Cu—1%–2%, S—6%, Au—200 µg g^{-1}, Pt—6 µg g^{-1}.

The mean efficiency of retention of Pt (500 ng) from sewage on 0.013 g of fungal biomass was about 60%. Therefore, higher biomass dosage (0.1 g) was used in further experiments. It was observed that the efficiency of biosorption of the same amount of Pt on higher biosorbent mass from sewage samples increased to 89% and from run-off samples achieved 95%. The recovery of Pt from digested fungal biomass was lower, in the range of 69%–82%, indicating that part of the Pt(IV) ions was non-specifically bound to the cells' surface. For 20 mL of standard and samples solutions containing Pt content in the range of 100–500 µg, the retention efficiency was in the range 95%–70%, while the recovery was in the range 70%–50%, decreasing with increasing Pt concentration (Table 3). The retention of analyte from samples containing 1000 µg of Pt decreased to about 57% from standard solutions and even below 50% for complex samples. In this case, the platinum uptake was only about 2.8 mg g^{-1}, not exceeding the maximal capacity of the sorbent. Probably the biosorption of high amounts of Pt was restricted by some factors. Competition between Pt and other ions, organic matter in sewage and road run-off samples, cannot be neglected.

Under the same conditions the recovery of Pd from biomass for all samples type was quantitative (Table 4). These results proved that Pd was specifically bound to fungal cells and no competition in biosorption was observed.

Owing to the presence of valuable metals and metalloids in the anode slime, products obtained from electrolytic refining of copper, numerous approaches have been made by the researchers to extract them following metallurgical routes [51]. The samples of anode slime were digested and pretreated according to the procedure described in Section 2.6. The content of Pt in anode slime determined by XRF method was 6 µg g^{-1}, while the content of Pd was 7 µg g^{-1}.

The experiment has shown that the recovery of precious metals from fungal biomass was excellent for Pd (97.9 ± 7.7%, n = 3) and good for Pt (74.8 ± 5.7%, n = 3). In this work, the biomass was digested by means of high temperature in concentrated acid. The obtained results show that the biomass of fungi can be applied for the quantitative recovery of palladium from wastewater and industrial samples. As for the recovery of precious metals from biomass an incineration process could be used [24], we also plan to examine the efficiency of this process in the future.

4. Conclusions

Yeast *Saccharomyces* sp. and fungi *Aspergillus* sp. have been used to investigate the biosorption of trace amounts of Pt(IV) and Pd(II) from chloride solutions in a batch system. The maximal efficiency of retention of Pt(IV) on yeast and fungi was obtained at pH 2.0 ± 0.2. The biosorption of Pd(II) was performed at pH 2.5 ± 0.2 on yeast, and pH 3.5 ± 0.2 on fungi. The biosorption process is very fast as the equilibrium was attained within 45 min. The best interpretation for the experimental data was given by the Langmuir isotherm and pseudo-second-order kinetics model. Fungi *Aspergillus* sp. show a higher adsorption capacity for both metals than yeast *Saccharomyces* sp. The maximum adsorption capacity of fungi was 5.49 mg g^{-1} for Pt(IV) and 4.28 mg g^{-1} for Pd(II). This microorganism was used for efficient removal of Pt and Pd from environmental samples (sewage, road run-off). The method could be also applied for the recovery of Pd from industrial samples (anode slime). The results revealed that, although this new biosorbent has a relatively low capacity, it is a promising candidate for removal of palladium from contaminated aquatic environments and industrial wastes. If the amount and value of metal recovered and if the biomass is plentiful, metal loaded biomass can be incinerated thereby eliminating further treatment.

Author Contributions: B.G.-Ż. conceptualization, writing, supervision, S.S. investigation, data calculation and preparation of figures, J.K. calculation of kinetics parameters.

Funding: This research received no external funding.

Conflicts of Interest: The authors declare no conflict of interest.

References

1. Bossi, T. Environmental Profile of Platinum Group Metals Interpretation of the results of a cradle-to-gate life cycle assessment of the production of pgms and the benefits of their use in a selected application. *Johns. Matthey Technol. Rev.* **2017**, *61*, 111–121. [CrossRef]
2. Cui, J.; Zhang, L. Metallurgical recovery of metals from electronic waste: A review. *J. Hazard. Mater.* **2008**, *158*, 228–256. [CrossRef] [PubMed]
3. Matthey, J. Precious Metal Division, Johnson Matthey Publishing Company. Available online: http://www.platinum.matthey.com/documents/new-item/pgm%20market%20reports/pgm-market-report-may-2016.pdf (accessed on 10 May 2019).
4. Zereini, F.; Alt, F. *Anthropogenic Platinum Group Element Emission*; Springer: Berlin, Germany, 2000.
5. Zereini, F.; Alt, F. *Palladium Emissions in the Environment, Analytical Methods, Enviromental Assessment and Health Effects*; Springer: Berlin, Germany, 2006.
6. Kalavrouziotis, I.K.; Koukoulakis, P.H. The environmental impact of the platinum group elements (Pt, Pd, Rh) emitted by the automobile catalyst converters. *Water Air Soil Pollut.* **2009**, *196*, 393–402. [CrossRef]
7. Jackson, M.T.; Richard, H.M.; Samson, J. Platinum-group elements in sewage sludge and incinerator ash in the United Kingdom: Assessment of PGE sources and mobility in cities. *Sci. Total Environ.* **2010**, *408*, 1276–1285. [CrossRef] [PubMed]
8. Rauch, S.; Morrison, G.M.; Motelica-Heino, M.; Donard, O.F.X.; Muris, M. Elemental association and fingerprints of traffic-related metals in road sediments. *Environ. Sci. Technol.* **2000**, *33*, 3119–3123. [CrossRef]
9. Nuss, P.; Blengini, G.A. Towards better monitoring of technology critical elements in Europe: Coupling of natural and anthropogenic cycles. *Sci. Total Environ.* **2018**, *613*, 569–578. [CrossRef] [PubMed]
10. Alhuwalia, S.S.; Goyal, D. Microbial and plant derived biomass for removal of heavy metals from wastewater. *Bioresour. Technol.* **2007**, *98*, 2243–2257. [CrossRef] [PubMed]
11. Park, D.; Yun, Y.S.; Park, J.M. The past, present and future trends of biosorption. *Biotechnol. Bioprocess Eng.* **2010**, *15*, 86–102. [CrossRef]
12. Das, N. Recovery of precious metals through biosorption–A review. *Hydrometallurgy* **2010**, *103*, 180–189. [CrossRef]
13. Lim, J.S.; Kim, S.M.; Lee, S.Y.; Stach, E.A.; Culver, J.N.; Harris, M.T. Quantitative study of Au(III) and Pd(II) ion biosorption on genetically engineered *Tobacco mosaic virus*. *J. Colloid Interface Sci.* **2010**, *342*, 455–461. [CrossRef]
14. Park, J.; Won, S.W.; Mao, J.; Kwak, I.S.; Yun, Y.S. Recovery of Pd(II) from hydrochloric solution using polyallylamine hydrochloride-modified *Escherichia coli* biomass. *J. Hazard. Mater.* **2010**, *181*, 794–800. [CrossRef] [PubMed]
15. Escudero, L.B.; Maniero, M.A.; Agostini, E.; Smichowski, P.N. Biological substrates: Green alternatives in trace elemental preconcentration and speciation analysis. *Trends Anal. Chem.* **2016**, *80*, 531–546. [CrossRef]
16. Chassary, P.; Vincent, T.; Marcano, J.S.; Macaskie, L.E.; Guibal, E. Palladium and platinum recovery from bicomponent mixtures using chitosan derivatives. *Hydrometallurgy* **2005**, *76*, 131–147. [CrossRef]
17. Ramesh, A.; Hasegawa, H.; Sugimoto, W.; Maki, T.; Ueda, K. Adsorption of gold(III), platinum(IV) and palladium(II) onto glycine modified crosslinked chitosan resin. *Bioresour. Technol.* **2008**, *99*, 3801–3809. [CrossRef] [PubMed]

18. Zhou, L.; Xu, J.; Liang, X.; Liu, Z. Adsorption of platinum(IV) and palladium(II) from aqueous solution by magnetic cross-linking chitosan nanoparticles modified with ethylenediamine. *J. Hazard. Mater.* **2010**, *182*, 518–524. [CrossRef] [PubMed]

19. Kim, Y.H.; Nakano, Y. Adsorption mechanism of palladium by redox within condensed-tannin gel. *Water Res.* **2005**, *39*, 1324–1330. [CrossRef] [PubMed]

20. Wang, R.; Liao, X.; Shi, B. Adsorption behaviors of Pt(II) and Pd(II) on collagen fibre immobilized bayberry tannin. *Ind. Eng. Chem. Res.* **2005**, *44*, 4221–4226. [CrossRef]

21. Sari, A.; Durali, M.; Tuzen, M.; Soylak, M. Biosorption of palladium(II) from aqueous solution by moss (*Racomitrium lanuginosum*) biomass: Equilibrium, kinetic and thermodynamic studies. *J. Hazard. Mater.* **2009**, *162*, 874–879. [CrossRef]

22. Parajuli, D.; Hirota, K. Recovery of palladium using chemically modified cedar wood powder. *J. Colloid Interface Sci.* **2009**, *338*, 371–375. [CrossRef]

23. Won, S.W.; Mao, J.; Kwak, I.S.; Sathishkumar, M.; Yun, Y. Platinum recovery from ICP wastewater by a combined method of biosorption and incineration. *Bioresour. Technol.* **2010**, *101*, 1135–1140. [CrossRef]

24. Won, S.W.; Lim, A.; Yun, Y.S. Recovery of high purity metallic Pd from Pd(II)-sorbed biosorbents by incineration. *Bioresour. Technol.* **2013**, *137*, 400–403. [CrossRef] [PubMed]

25. de Vargas, I.; Macaskie, L.E.; Guibal, E. Biosorption of palladium and platinum by sulfate-reducing bacteria. *J. Chem. Technol. Biotechnol.* **2004**, *79*, 49–56. [CrossRef]

26. Kim, S.; Song, M.H.; Wei, W.; Yun, Y.S. Selective biosorption behavior of *Escherichia coli* biomass toward Pd(II) in Pt(IV)-Pd(II) binary solution. *J. Hazard. Mater.* **2015**, *283*, 657–662. [CrossRef] [PubMed]

27. Xu, H.; Tan, L.; Dong, H.; He, J.; Liu, X.; Qiu, G.; He, Q.; Xie, J. Competitive biosorption behavior of Pt(IV) and Pd(II) by *Providencia vermicola*. *RSC Adv.* **2017**, *7*, 32229–32235. [CrossRef]

28. Turner, A.; Lewis, M.S.; Shams, L.; Brown, M.T. Uptake of platinum group elements by marine macroalga. *Mar. Chem.* **2007**, *105*, 271–280. [CrossRef]

29. Dziwulska, U.; Bajguz, A.; Godlewska-Żyłkiewicz, B. The use of algae *Chlorella vulgaris* immobilized on Cellex-T support for separation/preconcentration of trace amounts of platinum and palladium before GFAAS determination. *Anal. Lett.* **2004**, *37*, 2189–2203. [CrossRef]

30. Ju, X.; Igarashi, K.; Miyashita, S.I.; Mitsuhashi, H.; Inagaki, K.; Fujii, S.I.; Sawada, H.; Kuwabara, T.; Minoda, A. Effective and selective recovery of gold and palladium ions from metal wastewater using a sulfothermophilic red alga. *Galdieria sulphuraria. Bioresour. Technol.* **2016**, *211*, 759–764. [CrossRef] [PubMed]

31. Blackwell, K.J.; Singleton, I.; Tobin, J.M. Metal cation uptake by yeast: A review. *Appl. Microbiol. Biotechnol.* **1995**, *43*, 579–584. [CrossRef] [PubMed]

32. Mack, C.L.; Wilhelmi, B.; Duncan, J.R.; Burgess, J.E. A kinetic study of the recovery of platinum ions from an artificial solution by immobilized *Saccharomyces cerevisiae* biomass. *Miner. Eng.* **2008**, *21*, 31–37. [CrossRef]

33. Godlewska-Żyłkiewicz, B. Biosorption of platinum and palladium for their separation/ pre-concentration prior to graphite furnace atomic absorption spectrometric determination. *Spectrochim. Acta Part B* **2003**, *58*, 1531–1540. [CrossRef]

34. Godlewska-Żyłkiewicz, B.; Kozłowska, M. Solid phase extraction using immobilized yeast *Saccharomyces cerevisiae* for determination of palladium in road dust. *Anal. Chim. Acta* **2005**, *539*, 61–67. [CrossRef]

35. Malejko, J.; Szygałowicz, M.; Godlewska-Żyłkiewicz, B.; Kojło, A. Sorption of platinum on immobilized microorganisms for its on-line preconcentration and chemiluminescent determination in water samples. *Microchim Acta* **2012**, *176*, 429–435. [CrossRef] [PubMed]

36. Magnuson, J.; Lasure, L. Organic acid production by filamentous fungi. In *Advances in Fungal Biotechnology for Industry, Agriculture, and Medicine*; Tkacz, J., Lange, L., Eds.; Kluwer Academic & Plenum Publishers: New York, NY, USA, 2004; pp. 307–340.

37. Kapoor, A.; Viraraghavan, T. Heavy metal biosorption sites in *Aspergillus niger. Bioresour. Technol.* **1997**, *61*, 221–227. [CrossRef]

38. Akar, T.; Tunali, S. Biosorption characteristics of *Aspergillus flavus* biomass for removal of Pb (II) and Cu(II) ions from and aqueous solution. *Bioresour. Technol.* **2006**, *97*, 1780–1787. [CrossRef] [PubMed]

39. Baytak, S.; Kocyigit, A.; Turker, A.R. Determination of lead, iron and nickel in water and vegetable samples after preconcentration with *Aspergillus niger* loaded on silica gel. *Clean* **2007**, *35*, 607–611.

40. Moore, B.A.; Duncan, J.R.; Burgess, J.E. Fungal bioaccumulation of copper, nickel, gold and platinum. *Miner. Eng.* **2008**, *21*, 55–60. [CrossRef]

41. Woińska, S.; Godlewska-Żyłkiewicz, B. Determination of platinum and palladium in road dust after their separation on immobilized fungus by electrothemal atomic absorption spectrometry. *Spectrochim. Acta Part B* **2011**, *66*, 522–528. [CrossRef]

42. Sadowski, Z.; Maliszewska, I.H.; Grochowalska, B.; Polowczyk, I.; Koźlecki, T. Synthesis of silver nanoparticles using microorganisms. *Mater. Sci. Pol.* **2008**, *26*, 419–424.

43. Volesky, B. *Sorption and Biosorption*; Bv Sorbex: Montreal, QC, Canada, 2003; pp. 103–116.

44. Freundlich, H. Ueber die Adsorption in Loesungen. *Z. Physik. Chem.* **1907**, *57*, 385–470.

45. Langmuir, I. The adsorption of gases on plane surfaces of glass, mica and platinum. *J. Am. Chem. Soc.* **1918**, *40*, 1361–1403. [CrossRef]

46. Spieker, W.A.; Liu, J.; Miller, J.T.; Kropf, A.J.; Regalbuto, J.R. An EXAFS study of the co-ordination chemistry of hydrogen hexachloroplatinate(IV) 1. Speciation in aqueous solution. *Appl. Catal. A Gen.* **2002**, *232*, 219–235. [CrossRef]

47. Cabuk, A.; Akar, T.; Tunali, S.; Gedikli, S. Biosorption of Pb(II) by industrial strain of *Saccharomyces cerevisiae* immobilized on the biomatrix of cone biomass of *Pinus nigra*: Equilibrium and mechanism analysis. *Chem. Eng. J.* **2007**, *131*, 293–300. [CrossRef]

48. Sarri, S.; Misaelides, P.; Papanikolaou, M.; Zambulis, D. Uranium removal from acidic aqueous solutions by *Saccharomyces cerevisiae, Debaruomyces hansenii, Kluyveromyces marxianus and Candida colliculosa*. *J. Radioanal. Nucl. Chem.* **2009**, *279*, 709–711. [CrossRef]

49. Chen, C.; Wang, J. Influence of metal ionic characteristic on their biosorption capacity by *Saccharomyces cerevisiae*. *Appl. Microbiol. Biotechnol.* **2007**, *74*, 911–917. [CrossRef] [PubMed]

50. Netzahuatl-Muñoz, A.R.; del Carmen Cristiani-Urbina, M.; Cristiani-Urbina, E. Chromium biosorption from Cr(VI) aqueous solutions by *Cupressus lusitanica* bark: Kinetics, equilibrium an thermodynamic studies. *PLoS ONE* **2015**, *10*, e0137086. [CrossRef] [PubMed]

51. Hait, J.; Jana, R.K.; Sanyal, S.K. Processing of copper electrorefining anode slime: A review. *Miner. Process. Extr. Metall.* **2009**, *118*, 240–252. [CrossRef]

![water logo] *water*

MDPI

Article

Insights into the Kinetics of Intermediate Formation during Electrochemical Oxidation of the Organic Model Pollutant Salicylic Acid in Chloride Electrolyte

Noëmi Ambauen [1],*[iD], **Jens Muff** [2][iD], **Ngoc Lan Mai** [3], **Cynthia Hallé** [1], **Thuat T. Trinh** [1][iD] and **Thomas Meyn** [1][iD]

[1]   Department of Civil- and Environmental Engineering, Norwegian University of Science and Technology, 7491 Trondheim, Norway
[2]   Department of Chemistry and Bioscience, Aalborg University, 6700 Esbjerg, Denmark
[3]   Faculty of Applied Sciences, Ton Duc Thang University, Ho Chi Minh City, Vietnam
*   Correspondence: noemi.ambauen@ntnu.no; +47-7359-4748

Received: 28 May 2019; Accepted: 21 June 2019; Published: 26 June 2019

✓ check for updates

Abstract: The present study investigated the kinetics and formation of hydroxylated and chlorinated intermediates during electrochemical oxidation of salicylic acid (SA). A chloride (NaCl) and sulfate (Na_2SO_4) electrolyte were used, along with two different anode materials, boron doped diamond (BDD) and platinum (Pt). Bulk electrolysis of SA confirmed the formation of both hydroxylated and chlorinated intermediates. In line with the density functional theory (DFT) calculations performed in this study, 2,5- and 2,3-dihydroxybenzoic acid, 3- and 5- chlorosalicylic acid and 3,5-dichlorosalicylic acid were the dominating products. In the presence of a chloride electrolyte, the formation of chlorinated intermediates was the predominant oxidation mechanism on both BDD and Pt anodes. In the absence of a chloride electrolyte, hydroxylated intermediates prevailed on the Pt anode and suggested the formation of sulfonated SA intermediates on the BDD anode. Furthermore, direct oxidation at the anode surface only played a subordinate role. First order kinetic models successfully described the degradation of SA and the formation of the observed intermediates. Rate constants provided by the model showed that chlorination of SA can take place at up to more than 60 times faster rates than hydroxylation. In conclusion, the formation of chlorinated intermediates during electrochemical oxidation of the organic model pollutant SA is confirmed and found to be dominant in chloride containing waters.

Keywords: electrochemical oxidation; organic pollutant; salicylic acid; disinfection by-products; boron doped diamond; hydroxyl radicals; chlorinated intermediates; density functional theory

---

1. Introduction

Mono- and polycyclic aromatic compounds present in wastewaters such as landfill leachate, pose a special challenge when it comes to their removal from the water matrix. Greater parts of them are non-biodegradable, persist in the aquatic environment and demand a dedicated treatment step for their removal. Such persistent organic pollutants can be produced intentionally, such as pesticides, or are unintentionally produced during water disinfection [1]. They may enter the human body through the food chain via bioaccumulation [2]. Many persistent organic pollutants are suspected to be carcinogenic or have other detrimental effects on the aquatic environment [3]. Studies have shown that advanced oxidation processes (AOP) in general [4], and thereof electrochemical oxidation (EO) [5] specifically, are effective treatment processes for the removal of persistent organic pollutants. AOPs mainly focus on the formation of hydroxyl radicals by different means [6,7]. Hydroxyl radicals are non-selective, highly reactive oxidants and their presence leads to partial degradation or even complete oxidation to CO_2 of organic pollutants. During EO, adsorbed hydroxyl radicals are formed via the electrolytic discharge of water in the electrochemical cell (Equation (1)) [8]:

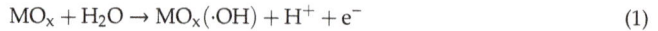

$$MO_x + H_2O \rightarrow MO_x(\cdot OH) + H^+ + e^- \tag{1}$$

A competing side reaction may occur during the electrolytic discharge of water, the so-called oxygen evolution (Equation (2)) [8]:

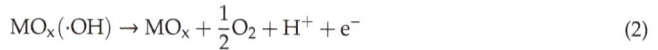

$$MO_x(\cdot OH) \rightarrow MO_x + \frac{1}{2}O_2 + H^+ + e^- \tag{2}$$

The occurrence of oxygen evolution is subject to the oxygen evolution over-potential of the anode material. In the same way, the transfer of oxygen from the adsorbed hydroxyl radicals to the organic compound (R) depends on the anode material [9]. Commonly, two different electrochemical oxygen transfer reaction (EOTR) mechanisms are distinguished [8]. The first EOTR mechanism takes place at anodes that have a low oxygen evolution over-potential, also known as active anodes (e.g., platinum, Pt) [8]. Hereby, the adsorbed hydroxyl radical reacts with the anode surface, forming higher oxides (MO_{x+1}), which in turn react with the organic pollutant (Equation (3)):

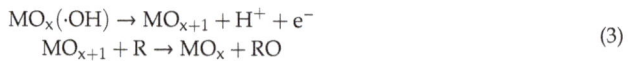

$$MO_x(\cdot OH) \rightarrow MO_{x+1} + H^+ + e^-$$
$$MO_{x+1} + R \rightarrow MO_x + RO \tag{3}$$

where MO_{x+1}/MO_x is the surface redox couple and also called active oxygen. The second EOTR mechanism takes place at anodes with high oxygen evolution over-potential, also known as non-active anodes (e.g., BDD). Hereby, no formation of higher oxides occurs. Instead, the adsorbed hydroxyl radicals react directly with the organic compound, which at best leads to its complete combustion to CO_2 (Equation (4)):

$$MO_x(\cdot OH) + R \rightarrow MO_x + mCO_2 + nH_2O + H^+ + e^- \tag{4}$$

Depending on the anode material, the active oxygen is called chemisorbed (active anodes) or physisorbed (non-active anodes). In order to be able to conduct EO, the sole presence of an organic pollutant and water molecules is not enough. Electrolytes need to be present, to transfer the charge across the electrochemical cell. Almost all wastewaters contain electrolytes of different natures and thus facilitate the treatment via EO. However, electrolytes do not only transfer charges, but electrochemically active electrolytes may also undergo oxidation at the anode surface. Halide electrolytes will form active halide species under such circumstances, for example NaCl will react to active chlorine. These active species will in turn react with the organic pollutants, resulting in their partial chemical oxidation, a process also known as mediated oxidation (MEO) [10]. Bonfatti et al. [11] proposed the following mechanisms for active chlorine mediated electrochemical oxidation:

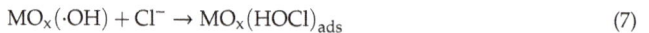

$$2Cl^- \rightarrow Cl_2 + 2e^- \tag{5}$$

$$Cl_2 + 2OH^- \rightarrow ClO^- + Cl^- + 2H^+ \leftrightarrow HOCl + H^+ + Cl^- \tag{6}$$

$$MO_x(\cdot OH) + Cl^- \rightarrow MO_x(HOCl)_{ads} \tag{7}$$

$$MOx(HOCl)_{ads} + R \rightarrow CO_2 + H_2O + Cl^- \tag{8}$$

Halogenated organic pollutants however are not favored reaction products, since they are generally more toxic than the mother compound and thus disadvantageous for the treatment process [12]. In the same way as the electrolytes, the organic pollutants can also be directly oxidized at the anodes surface, via a direct electron transfer (DET) between the organic pollutant and the anode [13]. DET is restricted and only occurs within the particular electrochemical window specific to a certain anode material, which in turn is limited by the above-mentioned oxygen evolution over-potential.

In this study, salicylic acid (SA) is used as a model compound for organic pollutants. SA is the main metabolite of acetylsalicylic acid, a commonly known and widely used painkiller. SA has aromatic properties and can be used to represent organic pollutants during electrochemical oxidation. Furthermore, chlorinated SA intermediates have been recently assigned to the group of disinfection by-products (DBP), which are not only of concern due to their cytotoxicity and growth inhibition ability but also because they act as a precursor to form regulated DBPs such as trihalomethanes (THMs). Consequently, this study investigates the degradation of SA in chloride and non-chloride electrolytes and investigates the different oxidation processes and the corresponding formation of intermediates.

Degradation pathways for SA during EO have been reported on previously. Guinea et al. [14] proposed a degradation pathway via hydroxylation of SA, using a BDD anode and cathodic generation of hydrogen peroxide. In the first oxidation step, SA reacted with the hydroxyl radicals originating from the hydrogen peroxide to form three different dihydroxybenzoic acids (dHBAs): 2,3 dihydroxybenzoic acid (23dHBA), 2,5 dihydroxybenzoic acid (25dHBA) and 2,6 dihydroxybenzoic acid (26dHBA). The different dHBAs were next proposed to be further oxidized to lower molecular weight carboxylic acids such as maleic or α-ketoglutaric acid. In a third oxidation step, the low molecular weight carboxylic acid was then degraded to oxalic acid and finally carbon dioxide (CO_2). Others suggested trihydroxybenzoic acids (tHBA), namely 234tHBA, 235tHBA and 246tHBA as possible degradation products of SA [15]. However, chlorinated products of SA formed during EO have not been reported. It is anticipated that chlorine atoms originating from the MEO primarily substitute at the para position of the hydroxyl group of SA, followed by a second substitution at the ortho position [16]. This assumption is endorsed by Broadwater et al. [17], where they identified different chlorinated SA intermediates by the simple chlorination of SA via the addition of NaOCl. Thus, it is expected that we will mainly find 3-chlorosalicylic acid (3ClSA) and 5-chlorosalicylic acid (5ClSA) and the combined product 3,5-dichlorosalicylic acid (35dClSA). The uncertainty that comes along with the expected chlorinated SA products can be reduced further by the implementation of preliminary density function theory (DFT) computations. These computations will help us to anticipate possible reaction products and thus facilitate the analytical process for their identification. The oxidation pathway of SA via DET is also included in this study as only a few studies investigated the DET of SA. Torriero et al. [18], used cyclic voltammetry (CV) to demonstrate irreversible DET of SA on a glassy carbon (GC) electrode, and Wudarska et al. [19] reported on the electro-reduction behavior of SA and acetylsalicylic acid during CV using a Pt electrode

The aim of this study is to gain more insight into the degradation of SA during EO with emphasis on the formation of chlorinated SA intermediates. Chlorinated intermediates, unlike their mother compound SA, belong to newly defined DBPs and therefore it is essential to elucidate their formation during EO and fill this gap of knowledge. Hydroxylated intermediates are also investigated, since they originate from the reaction with hydroxyl radicals, which are important for the removal of persistent organic pollutants. Kinetic models for the degradation of SA and the formation of intermediates are developed using DFT and tested through bulk electrolysis in different electrolytes. Model results provided rate constants that are used to assess the importance of different oxidation processes contributing to the degradation of SA. In addition, CV on BDD electrodes for SA are reported and add valuable information on the electroactive behavior of SA, which has been previously reported for different electrode materials and electrochemical reduction by [18,19], respectively.

2. Materials and Methods

Investigation of EOTR and MEO for SA was done by bulk electrolysis. Different mechanisms were identified via the reaction products of the parent compounds. The expected reaction products during bulk electrolysis were anticipated based on DFT computations for both EOTR and MEO. In addition, two different anode materials were tested, BDD and Pt, due to EOTR being highly dependent on the anode material. MEO is greatly affected by the supporting electrolyte, hence two different electrolytes were compared (NaCl and Na$_2$SO$_4$). A kinetic model has been developed in order to predict the degradation of SA and the formation of the reaction products during bulk electrolysis. Furthermore, CV was used to gain information about the electro-activity of SA. Its fate during CV was assessed by using different supporting electrolytes and anode materials. The CV results allow for conclusions to be drawn about the presence or absence of the DET mechanism during bulk electrolysis of SA.

2.1. Cyclic Voltammetry

SA and the two electrolytes (NaCl and Na$_2$SO$_4$) were purchased from VWR international AnalaR NORMAPUR. Stock solutions of SA were prepared with demineralized water with a concentration of 1 g/L. The stock solution was kept up to a month and stored in the dark at 4 °C. Electrolyte solutions were prepared with demineralized water. Cyclic voltammetry experiments were carried out with an μAUTOLABIII/FRA2 (Metrohm) using a 7.07×10^{-6} m^2 platinum rotating disc electrode (Pt-RDE, Metrohm), a 1.24×10^{-5} m^2 boron doped diamond rotating disc electrode (BDD-RDE, neoCoat) and a 1.19×10^{-4} m^2 glassy carbon rod (GC-rod) electrode (Metrohm AG, Hersiau, Switzerland). A Pt wire was used as an auxiliary electrode. The speed of the rotating disc electrode was set to 100 rpm for all experiments. A platinum wire served as a counter electrode and an Ag/AgCl (3M) was used as a reference electrode. All experiments were performed at room temperature (20 °C) using 25 mL electrolyte at a concentration of 0.1 M and a SA concentration of 500 mg/L. The scanning rate was 1 V/s and 5 consecutive scan cycles were run at the time and the potential was swept between −1 and 2 V. CV results were analyzed using NOVA 2.0 software (Metrohm).

2.2. Bulk Electrolysis

The same chemicals as for CV were used and stock solutions were prepared and stored likewise. Bulk electrolysis experiments were carried out, using a micro flow cell (ElectroCell Europe AS, Denmark). Experiments were conducted in galvanostatic mode with an applied current density (j$_{app}$) of 43 mA/cm^2. Two different anode materials were used, platinum (Pt) with titanium (Ti), as supporting material and a boron doped diamond (BDD) with niobium (Nb) as supporting material together with a stainless-steel cathode. Both, cathode and anode had an active area of 10 cm^2. Two polytetrafluoroethylene turbulence-enhancing meshes were placed between the anode and cathode (4 mm inter-electrode cap). The working- and counter-electrode were cooled during the experiment with a tab water stream (ca. 7 °C) from the rear side. The solution (2500 mL) in the tank was magnetically stirred and pumped with a peristaltic pump (Masterflex Cole-Parmer Instrument Co., Vernon Hills, IL, USA) via Teflon tubing to the micro flow cell with a flow rate of 380 mL/min. A cooling coil (stainless steel) immersed into to the solution and connected to a chiller (FP50-ME, Julabo GmbH, Seelbach, Germany) assured a stable temperature (25 °C) during all experiments. The electrolyte concentration used during bulk electrolysis was 0.05 M for both, NaCl and Na$_2$SO$_4$. Figure 1 depicts the scheme of the electrolysis set up.

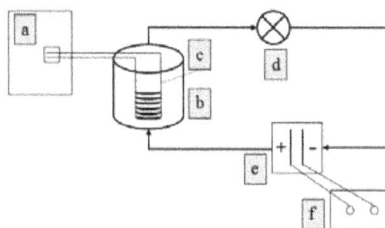

Figure 1. Scheme of electrolysis setup; a: chiller; b: tank; c: cooling coil; d: peristaltic pump; e: electrolytical cell; f: power supply.

The mass transfer coefficient (k_m) was obtained by the diffusion limiting current technique [20] according to Chatzismyeon et al. [21]. Different concentrations (4–24 mM) of potassium ferro cyanide ($K_4[Fe(CN)_6]$) and ferri cyanide ($K_3[Fe(CN)_6]$) in 2:1 ratio were anodically oxidized and polarization curves were generated. The limiting currents were determined by the formula:

$$I_{lim} = (AnFk_m)C_b \tag{9}$$

where I_{lim}: limiting current (A), A: electrode surface (m^2), n: number of exchanged electrons (n = 1 for ferro/ferri cyanide couple), F: Faraday constant (C/mol), k_m: mass transfer coefficient (m/s), C_b: bulk concentration of ferro cyanide (mol/m^3). The ratio of I_{lim} to C_b is obtained from the slope when plotting different ferro/ferri cyanide concentration versus the limiting current (plateau) from the polarization curves. The calculated k_m value is inserted into the following Equation (10), describing the initial limiting current density ($j_{lim}(t = 0)$) for the given experimental conditions when BDD anodes are used [22]:

$$j_{lim}(t) = nFk_mCOD(t) \tag{10}$$

where: n: number of electrons exchanged with anode (n = 4 when considering chemical oxygen demand (COD)), COD (t): initial bulk COD concentration at time t = 0).

2.3. Sample Analysis

Samples obtained from bulk electrolysis were analyzed using an UPLC (Waters, Milford, MA, USA) with XEVO TQ-XS triple quadrupole mass spectrometer (Waters) with a 2.1 mm × 100 mm high strength silica T3 column (Waters). The UPLC-MS/MS was operated in multiple reaction monitoring mode using electrospray ionization. Water (HPLC grade, VWR) with 2 mM ammonium formate (Sigma-Aldrich, Merck KGaA, Darmstadt, Germany) and 0.1% formic acid (VWR International LLC, Radnor, PA, USA) was used as solvent A and acetonitrile (HPLC grade, VWR) with 2 mM ammonium formate (Sigma-Aldrich) and 0.1% formic acid (VWR) was used as solvent B. For the SA method, a flow of 0.4 mL/min was constantly maintained and deuterated SA (SA-d6, Sigma-Aldrich) was used as an internal standard. Standards for the selection of expected SA products 23dHBA, 25dHBA, 26dHBA, 3ClSA, 4ClSA, 5ClSA and 35dClSA were purchased from Sigma-Aldrich, all analytical grade. Data processing was carried out using the 'Targetlynx' software (Waters).

Active chlorine in bulk electrolysis samples was measured with the DPD (N, N-diethyl-p-phenylenediamine) colorimetric method using DPD powder pillows for 5 mL (Hach, Loveland, CO, USA) and a portable DR300 colorimeter (Hach).

2.4. Density Functional Theory Simulations

DFT simulations were performed to study the relative stability of different products, as well as the electronic property of SA in reaction. All calculations were done with the Gaussian 09 package [23]. Unrestricted spin calculation using Lee-Yang-Parr (B3LYP) [24] functional and def2-TZVPP basis set [25,26] were employed. An implicit solvation model for water was considered using a solvation model based on the quantum mechanical charge density (SMD) [27]. Natural bond theory (NBO) [28] was used to analyze the spin and charge density of the molecules (Figure 2). The default values of Gaussian 09 were used for the convergence of energy and force in the DFT calculations. A similar set up was successfully employed to study to electro-chemical reaction [29].

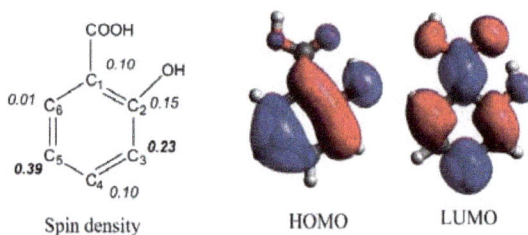

Spin density   HOMO   LUMO

Figure 2. Electronic structure of salicylic acid (SA) radical: ppin density calculated by natural bond theory (NBO) method (**left**) and highest occupied molecular orbital—lowest unoccupied molecular orbital (HOMO-LUMO) (**right**).

2.5. Kinetic Modelling

A mathematic model predicting the degradation of SA and its intermediates with different electrode materials was developed. A degradation mechanism of SA with NaCl as the supporting electrolyte is proposed in Figure 3. Note that in the case of using Na_2SO_4 as supporting electrolyte, the formation of chlorinated intermediates does not take place.

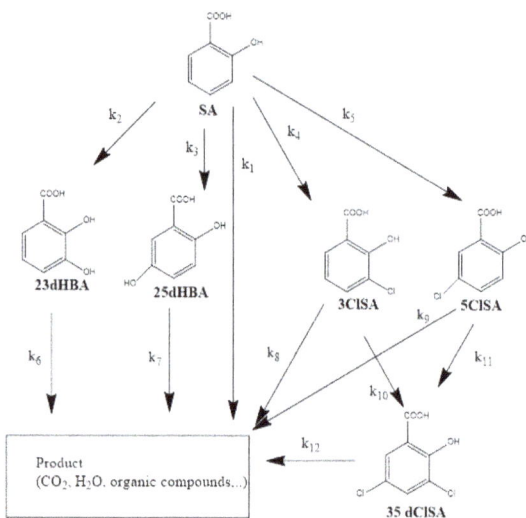

Figure 3. Simplified chemical kinetic model for the degradation of SA.

A first order kinetic equation was chosen to describe each reaction rate (Equation (11)). Calculations were performed in Matlab (version 2017, The MathWorks, Inc., Natick, MA, USA) to find the numerical solution to the set of ordinary differential equations. Reaction rate constants were determined by fitting the experimental data to the model using the least squares method. The fitting quality was estimated by the correlation coefficient R^2.

$$\frac{d[SA]}{dt} = -(k_1 + k_2 + k_3 + k_4 + k_5)[SA]$$
$$\frac{d[SA]}{dt} = -(k_1 + k_2 + k_3 + k_4 + k_5)[SA]$$
$$\frac{d[23dHBA]}{dt} = k_2[SA] - k_6[23dHBA]$$
$$\frac{d[25dHBA]}{dt} = k_3[SA] - k_7[25dHBA] \qquad (11)$$
$$\frac{d[3ClSA]}{dt} = k_4[SA] - (k_8 + k_{10})[3ClSA]$$
$$\frac{d[5ClSA]}{dt} = k_5[SA] - (k_9 + k_{11})[5ClSA]$$
$$\frac{d[35dClSA]}{dt} = k_{10}[3ClSA] + k_{11}[5ClSA] - k_{12}[35dClSA]$$

3. Results and Discussion

3.1. Prediction of Salicylic Acid Intermediate Formation by DFT Simulations

Considering the formation of hydroxylated and chlorinated products, there are six possibilities to bond to a carbon atom (from C1–C6) of SA (Figure 2). However, due to steric effects, the attacking position at C1 and C2 is very unstable, hence the possibility to bond with OH^- and Cl^- at position C3, C4, C5, and C6 was investigated. Table 1 presents the relative stability of the products, the most stable one (relative energy = 0.00 kJ/mol) was chosen as reference. For both the hydroxylated and mono-chlorinated products, the most stable structure is at position C4, while position 4,5 and position 3,5 are the most stable structures for di-chlorinated species. When looking at the experimental data, it shows that the most favorable attacking positions are at C3 and C5 position (discussed in Section 3.2.2 & 3.2.3). Thus, the most favorable intermediates observed during the experiments cannot be fully explained by the relative stability of products based on DFT calculations. It is therefore proposed that the mechanism and kinetics play an important role here. Consequently, further investigations of the electronic structure of the radical SA by DFT and NBO theory were performed. In particular, the spin density (Figure 2) shows that its maximum is at position C3 and C5. The spin density corresponds to the reactivity of the SA radical in the electrochemical oxidation reaction. Simply based on this assessment, it is possible to predict that the attraction of OH^- and Cl^- will be favorable at the C3 and C5 position. This is in agreement with the experimental data, where the byproducts for both hydroxylation and chlorination were observed at the C3 and C5 position.

Table 1. Relative energy of hydroxylated and chlorinated products.

Position Hydroxylate Product	Relative Energy (kJ/mol)	Position Mono-Chlorinated Product	Relative Energy (kJ/mol)	Position Di-Chlorinated product	Relative Energy (kJ/mol)
4	0.0	4	0.0	4,5	0.0
6	10.8	5	3.9	3,5	0.1
5	15.7	3	9.7	3,4	4.8
3	18.2	6	29.6	4,6	17.0
n/a	n/a	n/a	n/a	3,6	25.5
n/a	n/a	n/a	n/a	5,6	32.6

3.2. Oxidation of Salicylic Acid and Intermediate Formation

3.2.1. Electro-activity of Salicylic Acid

CV has been used to evaluate the electro-activity of SA. First, a GC rod was used as a baseline for the later comparison with different anode materials. GC is not prone to fouling, unlike the Pt electrode, yet it has an active surface, unlike BDD, meaning that compounds contained in the water matrix have a proper affinity to the anode surface. In a second step, the electro-activity of SA was assessed with a Pt and BDD electrode, which corresponded to the anode materials used for the bulk electrolysis experiments. CV using a GC anode revealed that SA is electro-active, as indicated by the anodic oxidation peak potential E_{pa} observed during the forward scan (Figure 4a).

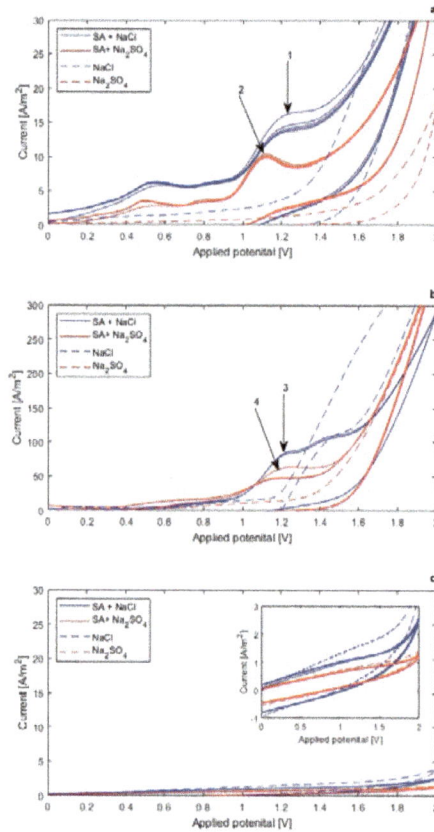

Figure 4. Cyclic voltammetry (CV) for SA in different supporting electrolytes and different anode materials (1V/s, 5 consecutive scan, current normalized for working electrode area). (**a**) GC, (**b**) Pt, (**c**) BDD.

No reversibility of the oxidation peak was observed during a reverse reduction scan of SA. This is in agreement with electrochemical irreversibility being a typical feature for phenolic compounds [30]. The use of two different supporting electrolytes, NaCl and Na_2SO_4, indicates that NaCl leads to a higher anodic peak current i_{pa} (Figure 4a, peak nr. 1) than Na_2SO_4 (Figure 4a, peak nr. 2) for SA. A higher anodic peak current for one supporting electrolyte indicates that the diffusion of the model compound towards the electrode surface is more efficient in NaCl than in Na_2SO_4, which can be associated to the different size of anions (Cl^- and SO_4^{2-}) [31]. Looking at the E_{pa} for GC and Pt anode, there is a slight shift towards a more positive potential for SA when NaCl is used instead of Na_2SO_4. The exact values for E_{pa} for each electrode material and supporting electrolyte are presented in Table 2.

Table 2. Overview of peak potentials E_{pa} (V) vs. Ag/AgCl (3M) for different anode materials and supporting electrolytes (scan rate: 1 V/s).

Anode	GC		Pt		BDD	
Electrolyte	NaCl	Na_2SO_4	NaCl	Na_2SO_4	NaCl	Na_2SO_4
E_{pa} (V) SA	1.20	1.07	1.22	1.15	n/a	n/a

In addition, SA exhibits a higher i_{pa} and a shift in E_{pa} in the first scan cycle compared to scan cycles 2–5 when using GC (Figure 4a) and Pt (Figure 4b, peak nr. 3 & 4) anodes. A lower E_{pa} in the second and more consecutive scan cycles can be attributed to the formation of a polymeric layer during the EO, which covers the electrode surface [32]. This results in a decreased i_{pa} for the consecutive scan cycles. Only a slight or no decrease of i_{pa} is observed after the second scan cycle which suggests that the polymeric layer is not developing any further. CV of SA on the BDD anode exhibited no oxidation peak E_{pa} (Figure 4c), which shows that SA does not undergo oxidation by DET. This is attributed to the non-active nature of BDD anodes, and results in a low affinity towards compounds contained in the water matrix. This behavior makes BDD anodes less prone to polymeric fouling than the active Pt anodes [8]. Contrary to our findings, Louhichi et al. [33] observed that SA is electro active on BDD electrodes during CV. They further state that a decreasing oxidation peak with increasing cycle numbers suggests a polymeric layer built up on the BDD anode. Furthermore, Montilla et al. [34] observed an anodic oxidation peak of the structurally similar benzoic acid on BDD anodes during CV. Differences in both studies include higher analyte concentrations, a different electrolyte (1M H_2SO_4 or 0.5 M $HClO_4$) and a considerably lower pH. In addition, the setup used in this study includes an RDE while it was not specified in the above-mentioned studies. Despite the fact that DET of SA on BDD electrodes was not observed in this study, BDD anodes are expected to outperform Pt anodes during bulk electrolysis of SA due to less electrode fouling and the formation of the more freely available physisorbed hydroxyl radicals [8].

CV results could verify the electro-activity of SA on active anodes (GC and Pt). The results confirm that electrons are directly exchanged between SA and these anodes, which contributes to the partial oxidation of SA. Thus, the use of different anode materials and electrolytes emphasized their impact on DET. Recorded anodic peak currents on Pt and GC electrodes showed that NaCl facilitated the transport of SA towards the electrode transport compared to Na_2SO_4.

3.2.2. Formation of Hydroxylated Salicylic Acid Intermediates

Bulk electrolysis was performed to investigate the oxidation of SA via EOTR. The original experimental data can be found in the supporting material (Tables S1–S4). With the NaCl-BDD setting, two hydroxylated products (23dHBA and 25dHBA) could be identified and quantified whereas the third expected hydroxylated product (26HBA) could not be detected at any time. Detected dHBAs were present after 10 min and they remained at a constant concentration throughout the experiment (Figure 5b). A concentration of 1.07×10^{-6} M and 4.85×10^{-6} M for 23dHBA and 25dHBA, respectively suggests a balanced formation and degradation during EO. Guinea et al. [14] also identified 25dHBA to be the most abundant among the three investigated dHBAs. A higher SA concentration (1.20×10^{-3} M) and cathodically generated hydrogen peroxide were used in that study, resulting in a significantly higher amount of each dHBA product. This explains why 26dHBA (limit of quantification (LOQ) of 10 nM) was not found in the present study but was found as a degradation product of SA by Guinea et al. [14].

When the NaCl-Pt setting was used, both, 23dHBA and 25dHBA could be identified and quantified, but 26dHBA could not be detected. In contrast to the BDD anode, the hydroxylated products do not show a constant production rate but do reach a maximum concentration after 60 min with 0.24×10^{-5} M and 0.69×10^{-5} M for 23dHBA and 25dHBA, respectively (Figure 5d). It is also notable that, after reaching a maximum concentration, both hydroxylated products are only eliminated again to a certain extent, similar to the observation on the BDD electrode. The final concentration was 0.13×10^{-5} M for 23dHBA and 0.56×10^{-5} M for 25dHBA. It should be noted that the absolute concentration of dHBAs detected was considerably higher than with the BDD anode, specifically 18% higher for 23dHBA and 13% higher for 25dHBA. This behavior can be attributed to the quasi freely available physisorbed hydroxyl radicals on BDD, which readily react with SA and lead to its complete mineralization (Equation (4)). However, on Pt anodes the hydroxyl radicals are chemisorbed (Equation (3)) and thus exhibit a lower oxidation power, which results in a lower amount of complete mineralized SA and a higher amount of the intermediate degradation products such as 23dHBA and 25dHBA. Similar findings have been reported by Madsen et al. [35].

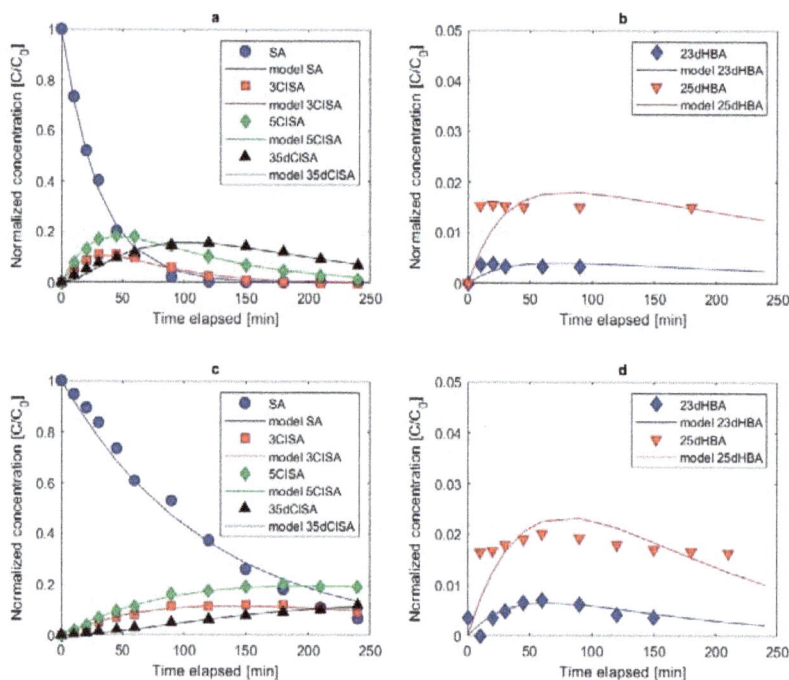

Figure 5. Experimental data and model with chloride (NaCl) as the supporting electrolyte: boron doped diamond (BDD) anode (**a**) and (**b**), Pt anode (**c**) and (**d**); normalized concentrations with respect to initial SA concentration.

Using the Na_2SO_4-BDD setting, 23dHBA and 25dHBA could also be identified and are depicted in Figure 6a,b, but 26dHBA was not detected. Both hydroxylated compounds were present as when NaCl was used. 23dHBA is present after 10 min at constant concentration of 1.3×10^{-6} M. 25dHBA shows the same pattern as 23dHBA with a concentration of 5.7×10^{-6} M.

With the Na_2SO_4-Pt setting, 23dHBA and 25dHBA could be observed yet 26dHBA could not be detected. However, the hydroxylated products differ in the concentration profile from the results obtained with the other three settings. A clear increase over time of both 23dHBA and 25dHBA is shown in Figure 6d with 3.24×10^{-6} M and 1.24×10^{-5} M as final concentrations for 23dHBA and 25dHBA, respectively.

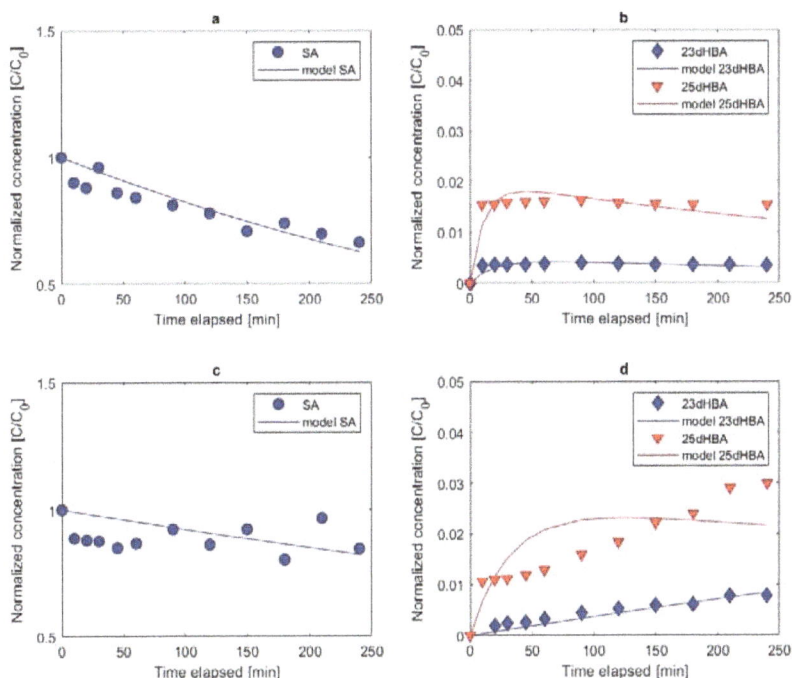

Figure 6. Experimental data and model with sulfate (Na$_2$SO$_4$) as the supporting electrolyte: BDD anode (**a**) and (**b**), Pt anode (**c**) and (**d**); normalized concentrations with respect to initial SA concentration.

3.2.3. Formation of Chlorinated Salicylic Acid

On the BDD anode with NaCl as the supporting electrolyte, the formation of 3ClSA, 5ClSA and 35dCLSA was observed instantly after the beginning of the experiment (Figure 5a). 3ClSA reached the maximum concentration (4.10 × 10^{-5} M) after 45 min and was further oxidized completely after 180 min. Likewise, 5ClSA reached its peak concentration (6.72 × 10^{-5} M) after 45 min and was almost completely oxidized (4.64 × 10^{-6} M) by the end of the experiment. 35dClSA reached its peak concentration (5.80 × 10^{-5} M) after 120 min and exhibited a final concentration of 2.56 × 10^{-5} M. The observed chlorinated salicylic acid products were formed as expected based on the previous DFT calculations and as suggested by Farinholt et al. [16] and Broadwater et al. [17].

Using a Pt anode showed the same chlorinated product formation i.e., 3ClSA, 5ClSA and 35dClSA (Figure 5c). All three identified chlorinated products were observed right after the beginning of the experiment. 3ClSA reached its peak concentration (4.17 × 10^{-5} M) after 150 min and decreased thereafter to a final concentration of 3.19 × 10^{-5} M. The peak concentration of 5ClSA (6.90 × 10^{-5} M) was reached after 180 min and decreased to a final concentration of 6.49 × 10^{-5} M. For 35DClSA a steady formation and no point of concentration inflection was reached until the end of the experiment where the final concentration amounted to 4.11 × 10^{-5} M.

Comparing the formation of the chlorinated products on BDD and Pt electrode (Figure 7) it shows that the maximum concentration of 3ClSA (BDD: 4.10 × 10^{-5} M, Pt: 4.18 × 10^{-5} M) and 5ClSA (BDD: 6.77 × 10^{-5} M, Pt: 6.90 × 10^{-5} M) are about equal. The LOQ corresponds to 1.50 × 10^{-7} M for both, 3ClSA and 5ClSA. For 35dClSA, a peak concentration (5.80 × 10^{-5} M) was only detected with the BDD electrode (Figure 5a). The experimental time of 240 min was too short to detect a peak concentration of 35dClSA with the Pt electrode and no apparent plateau was reached by the end of the experiment (Figure 5c).

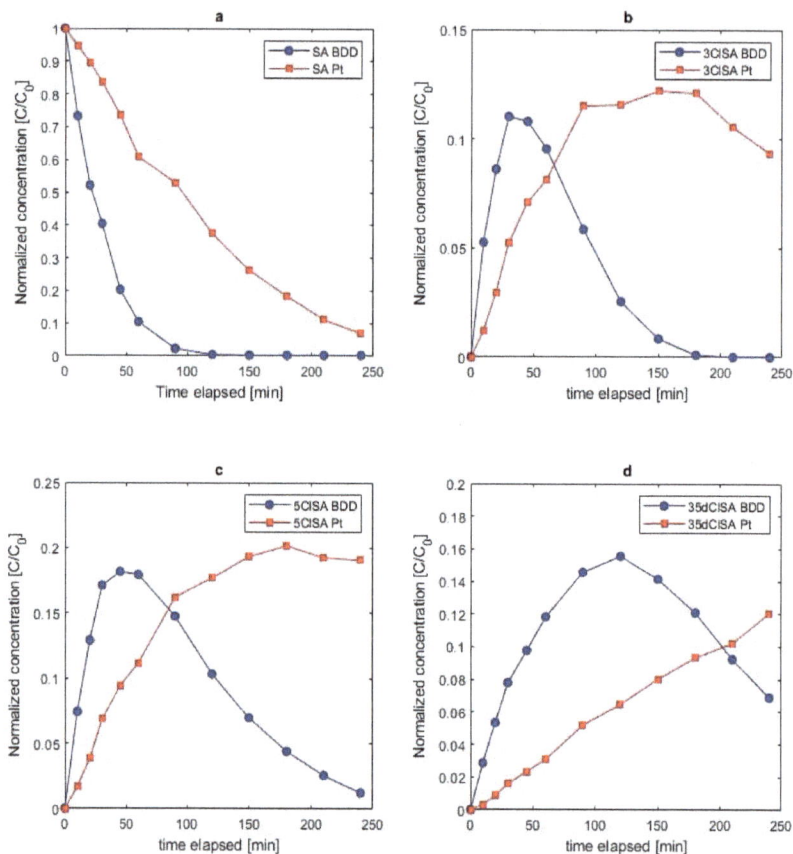

Figure 7. Normalized concentration profiles with respect to initial molar SA concentration [C0]: SA (**a**) chlorinated degradation products (**b–d**) on BDD and Pt electrodes.

3.2.4. Influence of Electrolyte Mediated Bulk Oxidation on Intermediate Formation

The two electrolytes form different reactive species that contribute to the MEO of SA. According to the reactivity of the oxidizing species formed, they also have a major impact on the degradation kinetics. The production of free chlorine was lower with the NaCl-Pt than with the NaCl-BDD setting. A final concentration of 9.3 mg Cl_2/L was measured with the Pt anode by the end of the experiment compared to 166 mg Cl_2/L with the BDD anode. This demonstrates that BDD not only favors the production of quasi-free hydroxyl radicals [8], but also favors the production of active chlorine Equations (7) and (8) and the consequent chlorine oxidation of organic compounds, which is the governing oxidation mechanism during EO when using NaCl as a supporting electrolyte. However, the similar production of hydroxylated products of SA on NaCl-BDD, NaCl-Pt and Na_2SO_4-BDD indicates that the hydroxylation takes part to the same extent regardless of electrolyte and anode material for the three mentioned parameter combinations. These findings show that in all three cases, hydroxylation competes with MEO since the formation of hydroxylated products does not increase. That leads to the further conclusion that despite the hydroxylation of SA there is also an MEO process, which governs the oxidation of SA on Na_2SO_4-BDD. This assumption is endorsed by the results of Farhat et al. [36] whereby they confirmed the important role of sulfate radicals formed on BDD anodes on the degradation of organic pollutants. Farhat et al. [36] suggested two possible pathways of sulfate radical formation on BDD anode, either via DET of SO_4^{2-} or by the reaction of H_2SO_4 or HSO_4^-

with the anodically generated hydroxyl radicals. Further, they suggest a nonradical activation of persulfate ($S_2O_8{}^{2-}$) that involves increasing the oxidation rate of organic contaminants. The Na_2SO_4-Pt parameter setting differs in its result compared to the other three parameter settings with regards to the concentration profile of the two hydroxylated compounds. The maximum concentration of both 23dHBA and 25dHBA differ by more than a factor 2 compared to the obtained concentration with the three other parameter settings. Since hydroxyl radical production is less on Pt than on BDD there is less mediating oxidizing species formed, as confirmed above concerning active chlorine species. Additionally, sulfate active species are not present when the Pt anode is used since the oxidation potential of sulfate (2.01 V vs. standard hydrogen electrode) exceeds the potential window of the Pt anode (depicted in Figure 4b) and oxygen evolution starts before sulfate oxidation. Thus, oxidation of SA at the Pt anode and Na_2SO_4 supporting electrolyte is governed by the higher oxides formed at active electrode surfaces, which are less reactive than the physically adsorbed hydroxyl radicals at the BDD anodes. This is reflected in slower overall degradation kinetics. Thus, the lack of oxidative species other than the higher oxides at the Pt anode surface explains why the partial oxidation of 23dHBA and 25dHBA is governing the process and not complete mineralization.

3.3. Overall Degradation Kinetics of Salicylic Acid

A degradation of 50 mg/L (362'004 nM) SA was followed over time for all four parameter settings and the apparent first order rate constants (k_{SA}) are presented in Table 3. Figure 5a depicts the results using BDD in combination with NaCl as the supporting electrolyte. SA could not be detected after 120 min (LOD of 30 nM). The kinetic behavior observed for SA was in accordance with the theory of the chosen mass transfer limiting conditions on BDD electrodes [22], where the applied current was higher than the limiting current (8 mA/cm^2). Oxidation of SA on Pt followed the same pattern as on the BDD anode and is depicted in Figure 5c. However, no complete degradation of SA was achieved during the experimental time whereby 7% of the initial SA concentration was still present at the end of the experiment. When Na_2SO_4 is used as the electrolyte the degradation of SA proceeds at a more moderate rate than with NaCl. For the Na_2SO_4-BDD setting, 66% of the initial SA concentration could still be observed at the end of the experiment (Figure 6a). The Na_2SO_4-Pt setting leads to a nearly constant concentration of SA with a final concentration corresponding to 84% of the initial SA concentration (Figure 6c).

The proposed chemical kinetic model (Figure 3) was fitted to the experimental data of SA, hydroxylated and chlorinated products and the corresponding computed results are depicted in Figures 5 and 6. In addition, Table 3 summarizes the rate constants (k) and the fitting quality (R^2) for the four different parameter settings.

The formation of hydroxylated products (k_2 and k_3) of SA are adequately described by the model and exhibit a R^2 between 77% and 88% for all four parameter settings. The only exception is 25dHBA with the Na_2SO_4-Pt setting and a R^2 of 64%, which means a less adequate fit of the model and the experimental data. This discrepancy can be explained by the scattered experimental data of SA and in consideration of the low amount of 25dHBA that is produced, i.e., less than 3% of SA is degraded to 25dHBA (Figure 6d). Because the kinetic model is assuming first order kinetics and the experimental data of 25dHBA does not exhibit clear first order kinetics, the fitting of the model to the experimental data is less accurate. Looking at the rate constants of the hydroxylated degradation products, it is evident that 25dHBA is formed faster (k_3) than 23dHBA (k_2) for all four parameter settings. These results are supported by the DFT calculation where the relative energy for 23dHBA was calculated to be higher than for 25dHBA, which means that 25dHBA is the more stable product. The degradation of the hydroxylated products to other organic products or complete combustion to CO_2 is expressed through k_6 and k_7 for 23dHBA and 25dHBA, respectively and is generally faster than their formation.

Table 3. Summary rate equations and fitting quality of the kinetic model.

Rate Constants (1/min)	NaCl/BDD	NaCl/Pt	Na$_2$SO$_4$/BDD	Na$_2$SO$_4$/Pt
k_1	1.28×10^{-2}	2.24×10^{-3}	3.32×10^{-14}	2.22×10^{-14}
k_2	1.75×10^{-4}	2.89×10^{-4}	2.41×10^{-4}	3.87×10^{-5}
k_3	7.62×10^{-4}	8.17×10^{-4}	1.71×10^{-3}	7.82×10^{-4}
k_4	8.56×10^{-3}	2.27×10^{-3}	n/a	n/a
k_5	1.10×10^{-2}	2.75×10^{-3}	n/a	n/a
k_6	3.63×10^{-3}	2.53×10^{-2}	5.07×10^{-2}	3.20×10^{-13}
k_7	2.89×10^{-3}	1.80×10^{-2}	8.70×10^{-2}	3.05×10^{-2}
k_8	2.22×10^{-14}	2.22×10^{-14}	n/a	n/a
k_9	9.81×10^{-3}	2.22×10^{-14}	n/a	n/a
k_{10}	2.59×10^{-2}	5.96×10^{-3}	n/a	n/a
k_{11}	3.55×10^{-3}	2.53×10^{-3}	n/a	n/a
k_{12}	9.96×10^{-3}	7.71×10^{-3}	n/a	n/a
$k_{SA}, \Sigma(k_1-k_5)$	3.33×10^{-2}	8.37×10^{-3}	1.95×10^{-3}	8.21×10^{-4}

Fitting Quality, R^2 (%)				
SA	98%	96%	95%	92%
23dHBA	77%	88%	87%	88%
25dHBA	78%	80%	88%	64%
3ClSA	94%	95%	n/a	n/a
5ClSA	96%	97%	n/a	n/a
35dClSA	93%	96%	n/a	n/a

Formation and degradation of chlorinated products of SA are described to an exceeding extent by the suggested kinetic model. R^2 for both, BDD and Pt anode are between 93% and 97%. The formation of 5ClSA proceeds at a faster rate (k_5) than the formation of 3ClSA (k_4) on both anode materials. Similar to the hydroxylation of SA, this can also be explained by the preliminary DFT calculations with the predicted relative energy for each compound (Table 1). The relative energy is lower for 5ClSA than for 3ClSA making 5ClSA the more stable product. The decrease of 3ClSA and 5ClSA consists of the formation of 35dClSA (k_{10} & k_{11}) and the degradation to other organic products or complete combustion to CO_2 (k_8 & k_9). The sums of k_8 and k_{10} for 3ClSA and k_9 and k_{10} for 5ClSA represent their degradation rate and show that 5ClSA is degraded at faster rate than 3ClSA on both anode materials. The degradation rate (k_{12}) of 35dClSA is slower than the one of the mono chlorinated SA on both, BDD and Pt anode.

When comparing the rate constants of hydroxylation and chlorination of SA it becomes evident that chlorination takes place at considerably faster rates than hydroxylation. On BDD anodes, the formation of 3ClSA is a factor 49 and 11 faster than the formation of 23dHBA and 25dHBA, respectively. Likewise, the formation of 5ClSA is a factor 63 and 14 faster than the formation of 23dHBA and 25dHBA, respectively. The same pattern can be observed for when the Pt anode is used yet the hydroxylation and chlorination rates differ less than with the BDD anode. 3ClSA is formed by a factor 8 and 3 faster than 23dHBA and 25dHBA, respectively. According to this, the formation of 5ClSA is a factor 10 and 3 faster than the formation of 23dHBA and 25dHBA, respectively. In conclusion, chlorination governs the degradation of SA during electrochemical oxidation, no matter the electrode material.

Finally, the kinetic model also provided the apparent rate constant of the SA degradation (k_{SA}), which consists of the sum of k_1 to k_5. Whereby k_1 accounts for the formation of other, not investigated organic products, the oxidation product of SA resulting from DET or other oxidation mechanisms of SA (Figure 3). Further, the degradation of SA to hydroxylated (k_2 and k_3) and chlorinated (k_4 and k_5) products are also accounted for in the apparent rate constant k_{SA}. The proposed kinetic model does not provide an individual rate constant for DET and thus its contribution to the degradation of SA is out of scope of this study. However, when NaCl is used instead of Na$_2$SO$_4$, k_1 is more than a factor 1.01×10^{11} higher for both anode materials and thus contributing to a higher degree to k_{SA} as when k_1 is obtained with Na$_2$SO$_4$. It indicates that if chlorination is not contributing to k_1 in terms of complete combustion of SA to CO2 or by the formation of other, not investigated chlorinated compounds the value of k_1 becomes significantly lower. This suggests that DET of SA, remains a mechanism that contributes to k_1 but plays a minor role in its degradation pathway. By comparing the different values for k_{SA} for the four parameter settings (Table 3) it becomes evident that the degradation of SA is in general faster with NaCl is used than with Na_2SO_4. However, among experiments using the same electrolyte, the use of a BDD anode resulted in a higher k_{SA} than with a Pt anode. On a BDD anode, the k_{SA} is a factor 17 higher for NaCl than for Na_2SO_4, and on a Pt anode the k_{SA} is a factor 10 higher for NaCl than for Na_2SO_4. Regardless of the parameter settings, the proposed kinetic model describes the experimental data of the SA degradation via k_{SA} to an exceeding extent with an R^2 of 92% or higher.

4. Conclusions

- This study confirms the formation of chlorinated intermediates. Three different chlorinated oxidation products were identified, 3ClSA, 5ClSA and 35dClSA, whereby 5ClSA was more frequently detected than 3ClSA and 35dClSA.
- Hydroxylation of salicylic acid via anodically generated hydroxyl radicals was confirmed via the identification and quantification of 23dHBA and 25dHBA. 25dHBA was more frequently formed than 23dHBA.
- Density functional theory and natural bond theory computations revealed the highest spin density at the C3 and C5 atom of salicylic acid. This explains the formation of the observed chlorinated and hydroxylated intermediates of salicylic acid, and why other intermediates like 26dHBA or 4ClSA were not detected.
- In chloride electrolyte, oxidation via mediating oxidizing species was found to be the governing oxidation process on both tested anode materials, whereas hydroxylation took place but at much lower rates than chlorination.
- Cyclic voltammetry confirmed of direct electron transfer of salicylic acid on Pt anodes, but not on BDD electrodes. The proposed kinetic model adequately describes the degradation of salicylic acid, and the formation of its chlorinated and hydroxylated intermediates and corresponding rate constants could be derived.

Supplementary Materials: The following are available online at http://www.mdpi.com/2073-4441/11/7/1322/s1, Tables S1–S4: original experimental data, Table S5: LOQ values of analysed compounds

Author Contributions: N.A., J.M., C.H. and T.M. conceived and designed the experiments. N.A. performed the experiments and analyzed the data. T.T and N.L.M. performed the DFT and NBT computations. T.T.T. wrote and validated the kinetic model. N.A. and T.T wrote the manuscript. J.M., C.H. and T.M reviewed and edited the manuscript. J.M., C.H. and T.M supervised the work.

Funding: This work is funded by the Norwegian University of Science and Technology (NTNU) and Søndre Helgeland Miljøverk IKS (SHMIL, Norway).

Acknowledgments: The authors acknowledge the computational resources of the Norwegian Metacentre for Computational Science (NOTUR). A special thank goes to Kåre Andre Kristiansen for his invaluable support in the MS-lab.

Conflicts of Interest: The authors declare no conflict of interest.

References

1. Fuoco, R.; Giannarelli, S. Integrity of aquatic ecosystems: An overview of a message from the South Pole on the level of persistent organic pollutants (POPs). *Microchem. J.* **2019**, *148*, 230–239. [CrossRef]
2. Li, Z. Health risk characterization of maximum legal exposures for persistent organic pollutant (POP) pesticides in residential soil: An analysis. *J. Environ. Manag.* **2018**, *205*, 163–173. [CrossRef] [PubMed]
3. Jones, K.C.; de Voogt, P. Persistent organic pollutants (POPs): State of the science. *Environ. Pollut.* **1999**, *100*, 209–221. [CrossRef]
4. Wang, F.; Smith, D.W.; El-Din, M.G. Application of advanced oxidation methods for landfill leachate treatment–A review. *J. Environ. Eng. Sci.* **2003**, *2*, 413–427. [CrossRef]

5. Panizza, M.; Delucchi, M.; Sirés, I. Electrochemical process for the treatment of landfill leachate. *J. Appl. Electrochem.* **2010**, *40*, 1721–1727. [CrossRef]

6. Andreozzi, R.; Caprio, V.; Insola, A.; Marotta, R. Advanced oxidation processes (AOP) for water purification and recovery. *Catal. Today* **1999**, *53*, 51–59. [CrossRef]

7. Dewil, R.; Mantzavinos, D.; Poulios, I.; Rodrigo, M.A. New perspectives for Advanced Oxidation Processes. *J. Environ. Manag.* **2017**, *195*, 93–99. [CrossRef]

8. Comninellis, C.; Chen, G. *Electrochemistry for the Environment*; Springer: New York, NY, USA, 2010. [CrossRef]

9. Foti, G.; Gandini, D.; Comninellis, C.; Perret, A.; Haenni, W. Oxidation of organics by intermediates of water discharge on IrO2 and synthetic diamond anodes. *Electrochem. Solid State Lett.* **1999**, *2*. [CrossRef]

10. Martínez-Huitle, C.A.; Ferro, S. Electrochemical oxidation of organic pollutants for the wastewater treatment: Direct and indirect processes. *Chem. Soc. Rev.* **2006**, *35*, 1324–1340. [CrossRef]

11. Bonfatti, F.; Ferro, S.; Lavezzo, F.; Malacarne, M.; Lodi, G.; de Battisti, A. Electrochemical Incineration of Glucose as a Model Organic Substrate II. Role of Active Chlorine Mediation. *J. Electrochem. Soc.* **2000**, *147*, 592–596. [CrossRef]

12. Stucki, S.; Kötz, R.; Carcer, B.; Suter, W. Electrochemical waste water treatment using high overvoltage anodes Part II: Anode performance and applications. *J. Appl. Electrochem.* **1991**, *21*, 99–104. [CrossRef]

13. Iniesta, J.; Michaud, P.; Panizza, M.; Cerisola, G.; Aldaz, A.; Comninellis, C. Electrochemical oxidation of phenol at boron-doped diamond electrode. *Electrochim. Acta* **2001**, *46*, 3573–3578. [CrossRef]

14. Guinea, E.; Arias, C.; Cabot, P.L.; Garrido, J.A.; Rodríguez, R.M.; Centellas, F.; Brillas, E. Mineralization of salicylic acid in acidic aqueous medium by electrochemical advanced oxidation processes using platinum and boron-doped diamond as anode and cathodically generated hydrogen peroxide. *Water Res.* **2008**, *42*, 499–511. [CrossRef] [PubMed]

15. Feng, L.; van Hullebusch, E.D.; Rodrigo, M.A.; Esposito, G.; Oturan, M.A. Removal of residual anti-inflammatory and analgesic pharmaceuticals from aqueous systems by electrochemical advanced oxidation processes, A. review. *Chem. Eng. J.* **2013**, *228*, 944–964. [CrossRef]

16. Farinholt, L.H.; Stuart, A.P.; Twiss, D. The Halogenation of Salicylic Acid. *J. Am. Chem. Soc.* **1940**, *62*, 1237–1241. [CrossRef]

17. Broadwater, M.A.; Swanson, T.L.; Sivey, J.D. Emerging investigators series: Comparing the inherent reactivity of often-overlooked aqueous chlorinating and brominating agents toward salicylic acid. *Environ. Sci. Water Res. Technol.* **2018**, *4*, 369–384. [CrossRef]

18. Torriero, A.A.J.; Luco, J.M.; Sereno, L.; Raba, J. Voltammetric determination of salicylic acid in pharmaceuticals formulations of acetylsalicylic acid. *Talanta* **2004**, *62*, 247–254. [CrossRef]

19. Wudarska, E.; Chrzescijanska, E.; Kusmierek, E. Electroreduction of Salicylic Acid, Acetylsalicylic Acid and Pharmaceutical Products Containing these Compounds. *Port. Electrochim. Acta* **2014**, *32*, 295–302. [CrossRef]

20. Wragg, A.A.; Tagg, D.J.; Patrick, M.A. Diffusion-controlled current distributions near cell entries and corners. *J. Appl. Electrochem.* **1980**, *10*, 43–47. [CrossRef]

21. Chatzisymeon, E.; Xekoukoulotakis, N.P.; Diamadopoulos, E.; Katsaounis, A.; Mantzavinos, D. Boron-doped diamond anodic treatment of olive mill wastewaters: Statistical analysis, kinetic modeling and biodegradability. *Water Res.* **2009**, *43*, 3999–4009. [CrossRef]

22. Panizza, M.; Kapalka, A.; Comninellis, C. Oxidation of organic pollutants on BDD anodes using modulated current electrolysis. *Electrochim. Acta* **2008**, *53*, 2289–2295. [CrossRef]

23. Frisch, D.J.F.M.J.; Trucks, G.W.; Schlegel, H.B.; Scuseria, G.E.; Robb, M.A.; Cheeseman, J.R.; Scalmani, G.; Barone, V.; Mennucci, B.; Petersson, G.A.; et al. Gaussian 09, revision B.01. 2010. Available online: http://gaussian.com/ (accessed on 13 October 2018).

24. Becke, A.D. Density-functional exchange-energy approximation with correct asymptotic behavior. *Phys. Rev. A* **1988**, *38*, 3098–3100. [CrossRef]

25. Weigend, F.; Furche, F.; Ahlrichs, R. Gaussian basis sets of quadruple zeta valence quality for atoms H-Kr. *J. Chem. Phys.* **2003**, *119*, 12753–12762. [CrossRef]

26. Weigend, F. Accurate Coulomb-fitting basis sets for H. to Rn. *Phys. Chem. Chem. Phys.* **2006**, *8*, 1057–1065. [CrossRef] [PubMed]

27. Marenich, A.V.; Cramer, C.J.; Truhlar, D.G. Universal Solvation Model Based on Solute Electron Density and on a Continuum Model of the Solvent Defined by the Bulk Dielectric Constant and Atomic Surface Tensions. *J. Phys. Chem. B* **2009**, *113*, 6378–6396. [CrossRef] [PubMed]

28. Foster, J.P.; Weinhold, F. Natural hybrid orbitals. *J. Am. Chem. Soc.* **1980**, *102*, 7211–7218. [CrossRef]

29. Jing, Y.; Chaplin, B.P. Mechanistic Study of the Validity of Using Hydroxyl Radical Probes To Characterize Electrochemical Advanced Oxidation Processes. *Environ. Sci. Technol.* **2017**, *51*, 2355–2365. [CrossRef]

30. Evans, D.; Hart, J.P.; Rees, G. Voltammetric Behaviour of Salicylic Acid at a Glassy Carbon Electrode and Its Determination in Serum Using Liquid Chromatography With Amperometric Detection. *Analyst* **1991**, *116*, 803–806. [CrossRef]

31. Dubois, D.; Moninot, G.; Kutner, W.; Jones, M.T.; Kadish, K.M. Electroreduction of Buckmlnsterfullerene, Electrolyte, and Temperature Effects Aprotic Solvents: Solvent Supporting. *J. Phys. Chem.* **1992**, 7137–7145. [CrossRef]

32. Lee, J.H.Q.; Koh, Y.R.; Webster, R.D. The electrochemical oxidation of diethylstilbestrol (DES) in acetonitrile. *J. Electroanal. Chem.* **2017**, *799*, 92–101. [CrossRef]

33. Louhichi, B.; Bensalash, N.; Gadri, A. Electrochemical oxidation of benzoic acid derivatives on boron doped diamond: Voltammetric study and galvanostatic electrolyses. *Chem. Eng. Technol.* **2006**, *29*, 944–950. [CrossRef]

34. Montilla, F.; Michaud, P.A.; Morallon, E.; Vazquez, J.L.; Comninellis, C.; Morallón, E.; Vázquez, J.L. Electrochemical oxidation of benzoic acid at boron-doped diamond electrodes. *Electrochim. Acta* **2002**, *47*, 3509–3513. [CrossRef]

35. Madsen, H.T.; Søgaard, E.G.; Muff, J. Chemosphere Study of degradation intermediates formed during electrochemical oxidation of pesticide residue 2,6-dichlorobenzamide (BAM) at boron doped diamond (BDD) and platinum–iridium anodes. *Chemosphere* **2014**, *109*, 84–91. [CrossRef] [PubMed]

36. Farhat, A.; Keller, J.; Tait, S.; Radjenovic, J. Removal of Persistent Organic Contaminants by Electrochemically Activated Sulfate. *Environ. Sci. Technol.* **2015**, *49*. [CrossRef] [PubMed]

water

MDPI

Article

Studies on the Kinetics of Doxazosin Degradation in Simulated Environmental Conditions and Selected Advanced Oxidation Processes

Joanna Karpinska *, Aneta Sokol, Jolanta Koldys and Artur Ratkiewicz

Department of Biology and Chemistry, University of Bialystok, Ciołkowskiego 1K, 15-245 Białystok, Poland; anetka_w@uwb.edu.pl (A.S.); jolantka-93@wp.pl (J.K.); artrat@uwb.edu.pl (A.R.)
* Correspondence: joasia@uwb.edu.pl

Received: 1 April 2019; Accepted: 10 May 2019; Published: 13 May 2019

check for
updates

Abstract: The photochemical behavior of doxazosin (DOX) in simulated environmental conditions using natural waters taken from local rivers as a solvent was studied. The chemical characteristics of applied waters was done and a correlation analysis was used to explain the impact of individual parameters of matrix on the rate of the DOX degradation. It was stated that DOX is a photoliable compound in an aqueous environment. Its degradation is promoted by basic medium, presence of environmentally important ions such as Cl^-, NO_3^-, SO_4^{2-} and organic matter. The kinetics of DOX reactions with OH^- and SO_4^- radicals were examined individually. The UV/H_2O_2, classical Fenton and photo-Fenton processes, were applied for the generation of hydroxyl radicals while the $UV/VIS:Fe_2(SO_4)_3:Na_2SO_2$ system was employed for production of SO_4^- radicals. The obtained results pointed that photo-Fenton, as well as $UV/VIS:Fe_2(SO_4)_3:Na_2SO_2$, are very reactive in ratio to DOX, leading to its complete degradation in a short time. A quantitative density functional theory (DFT) mechanistic study was carried out in order to explain the molecular mechanism of DOX degradation using the GAUSSIAN 09 program.

Keywords: doxazosin maleate; advanced oxidation processes; hydroxyl radical; sulfate radical; photodegradation; DFT study

---

1. Introduction

Recent environmental studies show an appearance of new atypical compounds in aquatic ecosystems on a global scale. Called Emerging Organic Contaminants (EOC), they are created by hundreds of organic compounds belonging to different chemical classes [1]. Some of them are natural components of an environment, making their presence detectable due to advancements in sample preparation procedures [2–4], as well as new detection techniques [5–7]. They have been detected in clean surface waters at few ng dm^{-3} levels while in polluted waters in the range from a few to hundreds of μg dm^{-3} [8,9]. Many EOC-s compounds do not cause acute toxicity, but their presence in the environment entails a number of adverse changes, including interference in animal as well as human endocrine systems [1]. Compounds that exhibit such activity or are suspected of it are named Endocrine Disrupting Compounds (EDC). According to the definition given by The Endocrine Society, EDCs are: "an exogenous chemical, or mixture of chemicals, that interferes with any aspect of hormone action" [10]. One of the more numerous groups belonging to EDCs are traces of pharmaceuticals [1,8]. The main identified source of pharmaceuticals in surface freshwater environments are from wastewater treatment plants (WTP) [11]. Some compounds from the EDC-s group, especially pharmaceuticals, possess biocidal properties or are resistant to biodegradation, so the WTP-s based on activated sludge technology are unable to remove all of them [12]. Therefore, the search for improvements that can be made to the water purification process is still currently a problem. The following modifications of the water technology were developed and introduced into practice: membrane bioreactors, purification ponds with aquatic plants, application of new microorganisms, enzymatic treatments [12] or advanced oxidation processes (AOPs) [13,14]. The advanced oxidation units are based on oxidation reactions of reactive chemical species such as hydroxyl, sulfate, chlorine and other radicals with organic pollutants [13,14]. Although AOPs are considered to be the most effective way for water treatment, their efficiency depends on many factors such as type of process, type and composition of the polluted water, and chemical properties of degraded contaminants [15,16]. Additionally, cost of operation as well as environmental implications should be considered [16]. Therefore, the introduction of AOP into the treatment process requires the optimization of the chemical conditions of an applied reaction and recognition of the chemical behavior of a main organic pollutant.

This paper presents the results of studies on kinetics of doxazosin (DOX) degradation under influence of light and some selected AOPs. Doxazosin mesylate [(4-amino-6,7-dimethoxy-2-quinazolinyl)-4-(1,4-benzodioxan-2-yl-carbonyl)-piperazine monomethansulphonate] (Figure 1) belongs to the group of (α_1)-adrenoreceptor antagonists [17]. It is used for the treatment of benign prostatic hyperplasia [17,18] and blood hypertension [18]. It is well absorbed by the digestive tract after oral administration. Afterwards, it is partially metabolized and excreted with urine in the unchanged form (about 4.8%) and in the form of metabolites: products of demethylation (23%) and hydroxylation (12%) [19]. To the best of our knowledge, DOX chemical behavior under the influence of light or AOPs has not been reported yet. Among many available AOPs reactions, the runs of UV/Vis direct photolysis, UV-H_2O_2, classical and photo-Fenton processes, and oxidation by $SO_4^{\cdot}$ were studied [16]. Attempts have been made to assess the persistence of DOX in aquatic environment and indicate the environmental factors affecting the rate of its vanishing. For this purpose, DOX degradation rate under irradiation of sunlight in the presence of natural matrix was determined.

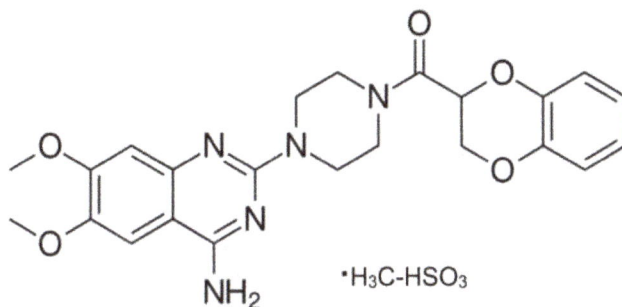

Figure 1. Chemical structure of doxazosin mesylate.

2. Materials and Methods

2.1. Chemicals

Doxazosin mesylate, DOX (Sigma-Aldrich, Germany), a stock solution at the concentration 2×10^{-3} mol dm^{-3} was prepared by dissolving an appropriate weight in 25 mL of MilliQ water. Working solutions at the concentrations 1.0×10^{-6}, 2.5×10^{-6}, 5.0×10^{-6}, 10^{-5}, 1.5×10^{-5} and 2×10^{-5} mol dm^{-3} were prepared by dilution in MilliQ water. Ferric sulfate (Fe$_2$(SO$_4$)$_3$) and anhydrous sodium sulfite (Na$_2$SO$_3$) were purchased from Chempur, Poland. Standard solution of ferric sulfate (2.5×10^{-3} mol dm^{-3}) was freshly prepared every day by the dissolution of an exact weighted amount in 50 mL of MilliQ water. Stock solution (5×10^{-3} mol dm^{-3}) of anhydrous sodium sulfite was freshly prepared every day from the pure product by dissolving an appropriate amount in 100 mL of MilliQ water.

Acetonitrile and methanol of HPLC grade were supplied by Sigma-Aldrich.

Tert–butyl alcohol (TBA) were purchased from Honeywell, Riedel-de Haën™.

Hydrogen peroxide (CHEMPUR, Poland) at the concentration of 10^{-1}, 5×10^{-2}, 10^{-2}, 2×10^{-2}, 2×10^{-3} and 10^{-3} mol dm^{-3} were prepared daily by suitably diluting its 30% solution in MilliQ water.

Other reagents used were: Concentrated acetic acid (Sigma-Aldrich, St. Louis, MO, USA), concentrated ammonium (Sigma-Aldrich), sodium hydroxide and sulfuric acid solutions at the concentration 1 mol dm^{-3} (POCh, Gliwice, Poland).

2.2. Irradiation Systems

All irradiation experiments were carried out using UV lamps and solar light simulator.

UV lamps—standard 16AV, (Cobrabid, Poznan, Poland) equipped with two light sources emitting radiation at 254 and 365 nm was used. All examined samples were irradiated by radiation at 365 nm as a representation of natural solar radiation UV-A.

Solar light simulator (SUNTEST CPS+, ATLAS, Champaign, IL, USA) emitting radiation in the range 300–800 nm was used for experiments in simulated natural conditions.

The intensity of light sources was measured using potassium Reinecke's salt actinometer. The intensity (Es) of radiation emitted by UV lamp was found to be 17.39 W m^{-2} while for the solar light simulator it was 19.53 W m^{-2}.

2.3. Absorbance Measurements

Monitoring of the current concentration of DOX was carried out spectrophotometically by reading the absorbance at 246 nm. For qualitative assessment of changes in DOX concentration, a calibration plot (ABS $= 5.2 \times 10^4 \pm 1.4 \times 10^2$ (DOX) $+ 0.8 \times 10^{-2} \pm 4.1 \times 10^{-3}$, $r^2 = 0.999$, where ABS—absorbance, (DOX) —concentration of DOX in mol dm^{-3}) was constructed for concentrations in the range 10^{-6}–2.0×10^{-5} mol dm^{-3}. The developed spectrophotometric method of DOX determination was characterized by low LOQ and LOD values equal 7.7×10^{-7} and 2.3×10^{-7} mol dm^{-3}, respectively. All spectrophotometric measurements were conducted with a Hitachi U-2800A spectrophotometer (Hitachi High-Technologies Europe GmbH (Mannheim Office), Mannheim, Germany). The following working settings of the device were used: scan speed 1200 nm min^{-1} and spectral bandwidth 1.5 nm.

2.4. Experimental Procedures

All irradiation experiments were conducted in a crystallization dish with 100 mL capacity with surface area open to atmosphere.

2.5. Direct Photolysis

50 milliliters of working solution of DOX at the concentration of 2.0×10^{-5} mol dm^{-3} was subjected to irradiation by a UV-lamp emitting radiation at 336 nm or to solar light in a solar simulator chamber. The spectrum of the solution was recorded every 10 min. A mixture of reagents without DOX irradiated at the same period of time was applied as a blank.

The pH of the aqueous solution was adjusted with 0.1 mol dm^{-3} H$_2$SO$_4$ or 0.1 mol dm^{-3} NaOH. pH was measured with an Elmetron CP-501 pH-meter (produced by ELMETRON, Zabrze, Poland) equipped with a pH-electrode EPS-1 (ELMETRON, Zabrze, Poland). The examination of photolysis in the environmental condition was done in the same manner as described above, using samples of surface water as a solvent.

2.6. H_2O_2—Assisted Photodegradation Process

H_2O_2-assisted photodegradation was studied using a working solution of DOX at the concentration 2.0×10^{-5} mol dm^{-3}. For this purpose, an appropriate volume of DOX aqueous solution was mixed with varying volumes of hydrogen peroxide so as to obtain final concentration of the oxidant in the range 10^{-1}–10^{-3} mol dm^{-3}. The pH of prepared mixtures was adjusted by adding a proper portion of NaOH or H_2SO_4 solution at the concentration 0.1 mol dm^{-3}. Mixtures prepared in this way were thereafter subjected to irradiation by UV lamp ($\lambda = 365$ nm) for 120 min. The spectrum of the reaction solution was recorded every 10 min using the irradiated mixture of reagents without DOX as a blank.

2.7. Fenton and Photo-Fenton Processes

The run of Fenton of photo-Fenton process was studied using an aqueous solution of DOX at concentration 2.0×10^{-5} mol dm^{-3}. For this purpose, a volume of 50 mL of working DOX solution acidified to an optimal pH by 0.1 mol dm^{-3} H_2SO_4 solution was mixed with variable volumes of H_2O_2 (10^{-2} mol dm^{-3}) and $FeSO_4$ (10^{-2} mol dm^{-3}). The molar ratio of Fenton reagent ingredients was kept 1:1, and their final concentrations were 10^{-4}, 5×10^{-5}, 10^{-5} and 5×10^{-6} mol dm^{-3}. Every 10 min, the spectrum of the reaction mixture was recorded against the mixture of reagents without DOX as a blank.

In the case of examination of the photo-Fenton process, the prepared mixtures were subjected to irradiation by UV light at 365 nm.

2.8. Photo Sulfite System

The following procedure was applied: initially, 0.456 mL of the doxazosin standard solution at the concentration of 2.0×10^{-3} mol dm^{-3} was introduced into a 50 mL volumetric flask. Next, a small volume of water was added followed by the introduction of 1 mL of ferric sulphate (VI) at the concentration of 2.5×10^{-3} mol dm^{-3} and 1 mL of sodium sulphite at the concentration of 0.05 mol dm^{-3}. After adding individual reagents, the 50 mL flask was filled to the mark with Milli-Q water. The prepared mixture was then subject to the irradiation by simulated solar light or UV light at $\lambda = 365$ nm. Like previously, the spectrum of the irradiated mixture was recorded using the irradiated mixture of reagents without DOX as a blank.

3. Results and Discussion

3.1. Initial Studies

At the beginning of the performed experiments, a UV spectrum of doxazosin aqueous solution was recorded. Its spectral characteristics possessed three distinct maxima: sharp and intense at 196 and 246 nm, and broad and less intense at 328 nm (Figure 2). The kinetics of doxazosin decay was observed by monitoring the changes at 246 nm.

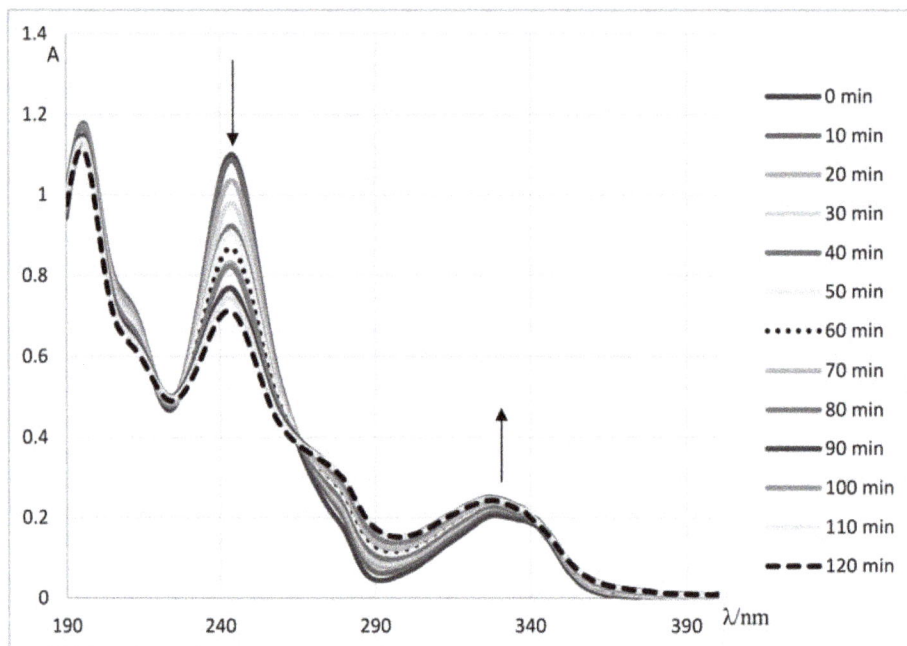

Figure 2. The changes in UV spectrum of aqueous doxazosin solution (c = 2.0 × 10^{-5} mol dm^{-3}) subjected to irradiation by simulated solar light at native pH 5.56 versus MilliQ water as a blank.

3.2. Direct Photolysis in Laboratory Conditions

The photostability of doxazosin in laboratory conditions was checked first. For this purpose, a portion of 50 mL of an aqueous solution of DOX at the concentration 2.0 × 10^{-5} mol dm^{-3} was subjected to irradiation by UV (λ = 365 nm) or simulated solar light. It was stated that DOX is a photoliable compound. The following changes in its spectral characteristics were observed: the intensity of the band at 246 nm was gradually decreased while at 328 nm was growing (Figure 2). The rate of the DOX photodegradation process is affected by the type of light, pH of the reaction solution, and the kind of accompanied matrix. It was observed that the process of direct photolysis is more evident under the influence of UV radiation. The obtained results are gathered in Table 1. Additionally, it was noticed that basic pH promotes DOX decomposition, presumably due to the hydrolysis process. Stability experiments were performed in order to confirm this assumption. For this purpose, a series of aqueous solutions of DOX with different pH (in range 1–13) were prepared and thermostated in 0, 30 and 80 °C for 6 h. The used test tubes were wrapped with aluminum foil in order to protect them against light. The UV spectrum of an examined solution was recorded every 20 minutes. It was stated that DOX is stable in an absence of stressed conditions in acidic, neutral and basic medium [20]. Slight signs of hydrolysis were observed in basic solutions heated at the temperature of 80 °C. The assayed value of the hydrolysis rate constant at a temperature of 80 °C and in a basic medium was equal to 3 × 10^{-5} min^{-1}.

Table 1. The kinetics parameters of doxazosin photodegradation in laboratory solutions and in the presence of natural matrix.

Studied Process	Used Irradiation		pH	k/min^{-1}	$t_{1/2}/min$	% of Degradation
Direct photolysis	UV 365 nm		9	2.2×10^{-3}	314	24
	UV 254 nm		5.56	7×10^{-4}	986	13
			9.0	4.2×10^{-3}	164	53
	Suntest		5.56	4.3×10^{-3} (0–60 min) 3.0×10^{-3} (61–120 min)	160 (0–60 min) 230 (61–120 min)	46
			9.0	1.6×10^{-2} (0–60 min) 1.7×10^{-3} (61–120 min)	44 (0–60 min) 406 (61–120 min)	71
	River I	UV_{365} nm	7.94	2.60×10^{-3}	266	25
		Suntest		7.0×10^{-3}	98	54
	River II	UV_{365} nm	8.23	2.80×10^{-3}	248	27
		Suntest		8.5×10^{-3}	82	61
Direct photolysis in presence of natural matrix	River III	UV_{365} nm	7.54	2.4×10^{-3}	290	23
		Suntest		14.5×10^{-3}	48	80
	River IV	UV_{365} nm	7.29	3.1×10^{-3}	223	29
		Suntest		11.0×10^{-3}	64	70
	Carbonate buffer	Suntest	8.3	8.0×10^{-3}	87	59

3.3. Factors Influencing Photolysis of Doxazosin in Environmental Conditions

The stability of doxazosin under simulated environmental conditions was checked next. The goal of this experiment was to answer what the persistence of this compound in natural conditions was. As the chemical composition of natural surface waters is very complex and difficult for reconstruction, real samples of water taken from local rivers were used as solvents for preparing working solutions of DOX. The chemical assessment of the quality of the applied waters showed that the rivers from which the samples were supplied were unpolluted (Table 2). Only in the case of river 3 did the levels of SO_4^{2-} and NO_3^- ions exceed the acceptable reference values. This river flows through agricultural areas and the elevated levels of these ions may be caused by run-offs of fertilizers from fields.

Table 2. Chemical characteristics of the used waters.

Parameter	River 1 53°7′ N; 23°7′ E	River 2 53°29′ N; 22°44′ E	River 3 52°20′ N; 23°03′ E	River 4 52°57′ N; 22°57′ E	Reference Value	Ref.
pH	7.94	8.23	7.54	7.29	3–11	44, 45
Conductivity/μS/cm	530	560	330	460	10–4000	46
SO_4^{2-}/mg L^{-1}	15.16	77.33	116.40	14.10	10–80	47
NO_3^-/mg L^{-1}	70.00	22.84	21.88	35.58	<50	44, 45, 48
Cl$^-$/mg L^{-1}	41.40	10.70	199.00	35.50	0.4–170	49
HCO_3^-/mval L^{-1}	5.80	5.00	4.80	5.60	<14	45
Ca/mg L^{-1}	101.70	9.29	9.29	75.80	<250	50
Mg/mg L^{-1}	5.98	2.82	2.57	6.20	<150	50
Fe$_{diss}$/mg L^{-1}	0.33	0.23	0.04	0.77	<2	51
TOC (total organic carbon)/mg L^{-1}	4.40	1.74	1.69	1.62	<40	52
$O_{2(diss)}$/mg L^{-1}	10.88	54.70	37.30	15.40	>4	53

The photochemical experiments proved that natural waters created an effective chemical system [21]. The observed decomposition rates were similar and strongly dependent on the kind of irradiation used. It was observed that rates of the degradation of DOX under influence of solar light ran two to five times faster than those of the UV-induced process (Table 1) and varied in the range of 7.0×10^{-3}–14.5×10^{-3} min^{-1}. The photolysis experiments with laboratory solutions of DOX in the presence of carbonate ions (pH 8.3) implied that DOX is photoliable compounds and its photodecomposition proceeds mainly via direct photolysis. The created intermediate products, radicals, were initiating a chain process which is inhibited in the presence of radical quenchers such as organic matter or carbonate ions. It was observed that the presence of carbonate and bicarbonate ions did not affect the rate of studied process. This observation allowed us to conclude that the degradation of DOX was independent of changes in the concentration of free radicals in the irradiated solution, but a situation in natural environment was more complicated. This medium is rich in organic matter and a variety of inorganic ions. It is known that dissolved organic matter is photoliable and its products launch a series of reactions with accompanied chemical species [22,23]. If the decomposition of DOX occurs as a result of direct photolysis alone, the rate of its decomposition in the presence of such complex matrix should decrease due to competition for light access. The observed rate of DOX degradation was at least twice higher than this for laboratory solutions. Not so high acceleration of the degradation rate in the presence of the matrix from river 1 could be attributed to a decrease in an energy flux attained by the doxazosin molecules. The acceleration of the decomposition rate, especially visible in the water from river 3, was probably caused by the presence of a variety of inorganic ions which are photosensitive. The photochemical reactions of chloride, nitrate and sulfate ions created a complicated chain of radical reactions which led to the generation of hydrogen peroxide and hydroxyl, sulfate and nitrate radicals, as well as other radicals [24–27]. The high rate of DOX disappearance in the presence of matrix from the river 3 can be explained as a synergistic action of the system rich in reactive species derived from inorganic ions and organic matter. Chloride radicals can be generated as a result of the direct photolysis:

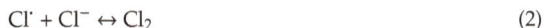

$$Cl^- + h\nu \rightarrow Cl^. + e^- \tag{1}$$

$$Cl^. + Cl^- \leftrightarrow Cl_2 \tag{2}$$

or as the result of their interaction with other oxidants, among other excited triplet states of organic sensitizers (^{3}SENS*) [25]:

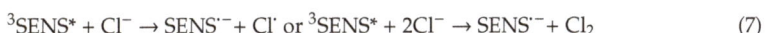

$$OH^. + Cl^- \leftrightarrow HOCl^- \leftrightarrow OH^- + Cl \tag{3}$$

$$HOCl^{.-} \leftrightarrow H^+ \leftrightarrow H_2O + Cl \tag{4}$$

$$HOCl^{.-} + Cl^- \leftrightarrow O^- + Cl_2 \tag{5}$$

$$SO_4^. \rightarrow SO_4^{2-} + Cl \tag{6}$$

$$^3SENS^* + Cl^- \rightarrow SENS^{.-} + Cl^. \text{ or } ^3SENS^* + 2Cl^- \rightarrow SENS^{.-} + Cl_2 \tag{7}$$

The photochemical reactions of NO_3^- contribute to an increase of the overall concentration of reactive species in the reaction environment as a consequence of the following processes [27,28]:

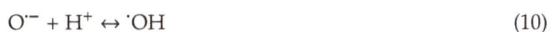

$$NO_3^- + h\nu \rightarrow NO_3^{-*} \rightarrow NO_2^- + O(^3P) \tag{8}$$

$$\text{or } NO_3^- + h\nu \rightarrow NO_3^{-*} \rightarrow NO_2^. + O^{.-} \tag{9}$$

$$O^{.-} + H^+ \leftrightarrow ^.OH \tag{10}$$

The following equilibria are established in the presence of sulfate ions [29]:

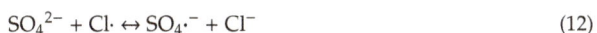

$$H^+ + SO_4^{2-} \leftrightarrow HSO_4^- \tag{11}$$

$$SO_4^{2-} + Cl^. \leftrightarrow SO_4^{.-} + Cl^- \tag{12}$$

The HSO_4^- ion reacts with $OH^.$ radical producing less reactive sulfate radical which however, is involved in the production of more reactive species [29]:

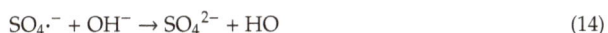

$$SO_4^{.-} + H_2O \rightarrow H^+ + SO_4^{2-} + HO \tag{13}$$

$$SO_4^{.-} + OH^- \rightarrow SO_4^{2-} + HO \tag{14}$$

The acceleration of the decomposition rate of the DOX solution with matrix from river 4 can be assigned to the photo-Fenton process occurring in the presence of dissolved organic matter which is responsible for production of the reactive radicals (HO, $O(^3P)$, O^-, H_2O_2) [30,31].

The obtained results pointed out that the decomposition of doxazosin in a natural environment is a very complex process that depends on the chemical composition of an accompanied matrix. It can be stated that the photoreactions of the matrix lead to the increase of the overall concentration of highly reactive radicals such as $HO^.$, which is predominant and mainly responsible for the acceleration of DOX decomposition rate [32].

3.4. Kinetics of DOX Decomposition Under Influence of Some Advanced Oxidation Processes

Four advanced oxidation systems: UV/H_2O_2, classical, and photocatalytic Fenton reaction, and photo-sulfite systems were chosen from among a number of possible AOPs methods, and their efficiency in DOX degradation were examined. The kinetics of DOX decomposition under the influence of UV/H_2O_2, system was studied first. The influence of an oxidant concentration in the range 10^{-4}–10^{-2}

mol dm^{-3} and pH (3.85–8) was checked. It was stated that the degradation of DOX in the UV/H$_2$O$_2$ system fit the pseudo-first order reaction. An addition of hydrogen peroxide caused a three-to-eight-fold increase in the reaction rate in comparison to the direct photolysis process. The observed enhancement depends on the used light, concentration of oxidant, and pH (Tables 1 and 3). It was noted that the increase in hydrogen peroxide concentration increased the reaction rate, but this augmentation was rather slight. The 100-fold reinforcement in the oxidant concentration amplified the reaction rate by only 7%. Analogically, as in the case of direct photolysis, the basic pH promoted the studied process, as the acidic pH of the rate of reaction was almost negligible. The role of the oxidant and light was checked next. For this purpose, two series of DOX aqueous solutions at pH 8 with the concentrations 2.0×10^{-5} mol dm^{-3} and 2.5×10^{-5} mol dm^{-3} were prepared. Appropriate volumes of H$_2$O$_2$ working solutions were added to each test tube so that the concentration of the oxidant was in the range 10^{-5}–5×10^{-4} mol dm^{-3}. Each test tube was wrapped with aluminum foil in order to protect against light and subsequently left at ambient temperature for 24 h. Thereafter, the spectra of each mixture were recorded. No changes in the spectral characteristics of doxazosin were observed, which proves the important dual role of light in this process. On the one hand, light induces the process of direct DOX photolysis, while on the other hand it breaks down the dihydrogen peroxide into hydroxyl radicals according to the reaction [33]:

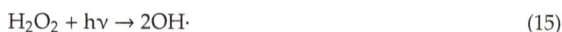

$$H_2O_2 + h\nu \rightarrow 2OH\cdot \tag{15}$$

Considering the above findings, it could be concluded that the observed enhancement in the rate of DOX degradation is a sum of the two above processes.

The classical Fenton system consisting of a solution of inorganic ferrous salt and hydrogen peroxide was examined next. The operation of the Fenton system is very complex and not fully known yet. The reaction mechanism involves the generation of hydroxyl radical according to the following reaction) [34]:

$$Fe^{2+} + H_2O_2 \rightarrow Fe^{3+} + OH\cdot + OH^- \tag{16}$$

Its efficiency in the degradation of organic pollutants is affected by the concentration of reagents, their molar ratio, and pH of reaction medium. In order to select the optimal concentrations of Fenton reagent components allowed to follow DOX degradation kinetics, a series of experiments were carried out using different concentrations of ingredients at their molar ratio of 1: 1. For this purpose, the concentrations in the range 5×10^{-6}–5×10^{-4} mol dm^{-3} were applied. The second order kinetics was assumed for studied Fenton and photo-Fenton systems. It was stated that at the lowest examined concentration, the observed process proceeded too slowly, but when the highest one was used, the total disappearance of the drug was observed in five minutes after the initiation of the reaction. The Fenton reagent with the concentration of 10^{-4} mol dm^{-3} of the components was selected for further testing, resulting in the determination of the kinetic parameters of DOX degradation with good precision. The influence of pH was checked next; it is known that the optimal working pH for the studied process is contained in the range 2–5 [34]. At an excess of hydrogen ions, less reactive positively charged ferrous species are formed [35]. Additionally, the surplus of hydrogen ions act as radical scavengers according to the following reaction:

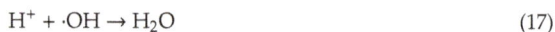

$$H^+ + \cdot OH \rightarrow H_2O \tag{17}$$

The precipitation of Fe(II) and Fe(III) hydroxides is observed at pH > 5. In order to select the best pH for the DOX degradation, rates of reaction at various pH values in the range 2–5 were measured. The obtained output showed that pH 3.5 was optimal for the studied process.

The influence of molar ratio of the Fenton reagent constituents on kinetics of DOX degradation was checked. The following molar ratios of $n_{H2O2}:n_{Fe(III)}$ = 1:1, 2:1, 10:1, and 1:2, 1:5, 1:10 were used. The obtained results demonstrated (Table 3) that the use of an excess of hydrogen peroxide in ratio to ferrous ions promotes the degradation process while the reverse ratio is unfavorable for the course of the reaction. Following this, the effect of UV radiation on the course of the DOX decomposition reaction with the Fenton reagent was examined. The results shown in Table 3 proved that application of light enhanced the efficiency of the Fenton process due to the following process [34]:

$$Fe^{3+} + h\nu + H_2O \rightarrow Fe^{2+} + \cdot OH + H^+ \tag{18}$$

The regeneration of ferrous ions increased the overall concentration of hydroxide radicals and slowed down an increase in pH of reaction medium [34].

Recently, the use of sulphate radicals to remove organic compounds from waters has attracted more attention due to their stability and high oxidation potential (2.5–3.1 V vs. NHE) [36]. Additionally, they can work in a wide range of pH, so a rigid control of this parameter is not necessary [37,38]. The prevailing type of oxidant in the reaction environment depends on value of the initial pH. It was found that $SO_4\cdot^-$ radicals predominate at the acidic pH, while at the basic pH $\cdot OH$ radicals are responsible for oxidation of organic compounds [37,38]. The only disadvantage of using sulphate radicals is the necessity to activate precursors to obtain the right concentration of the oxidant [37,38]. An alternative to conventional methods for production of $SO_4\cdot^-$ radicals ferric sulphate-sodium sulphite system in the presence of light was proposed [39]. This system is based on Fe-catalyzed sulphite oxidation and photochemical cycle of Fe(III)-Fe(II) species. For this reason, it can be considered a modification of the Fenton system [39]:

$$Fe^{3+} + HSO_3^- \leftrightarrow FeSO_3^+ + H^+ \tag{19}$$

$$FeSO_3^+ \rightarrow Fe^{2+} + SO_3\cdot^- \tag{20}$$

$$SO_3\cdot^- + O_2 \rightarrow SO_5\cdot^- \tag{21}$$

$$SO_5\cdot^- + HSO_3^- \rightarrow SO_4\cdot^- + SO_4^{2-} \tag{22}$$

$$SO_5\cdot^- + SO_5\cdot^- \leftrightarrow 2SO_4\cdot^- + O_2 \tag{23}$$

$$SO_5\cdot^- + SO_5\cdot^- \leftrightarrow SO_3\cdot^- + HSO_5^- \tag{24}$$

$$Fe^{2+} + HSO_5^- \rightarrow SO_4\cdot^- + Fe^{3+} + OH^- \tag{25}$$

$$FeSO_3^+ + light \rightarrow Fe^{2+} + SO_3\cdot^- \tag{26}$$

$$FeOH^{2+} + light \rightarrow Fe^{2+} + \cdot OH \tag{27}$$

The results of the above chain of reactions is a mixture of a variety of radicals where $SO_4\cdot^-$ and $\cdot OH$ are predominant [39].

The kinetics of DOX degradation under the influence of UV/Vis-Fe(III)-sulphite system was examined. For this purpose, a series of DOX solutions at concentration 2.0×10^{-5} mol dm^{-3} were mixed with variable volumes of ferric sulphate solution at the concentration 2.5×10^{-3} mol dm^{-3} and sodium sulphite at the concentration 5×10^{-2} mol dm^{-3}. The applied concentrations of reagents are shown in Table 3.

The obtained results showed that the efficiency of light- Fe(III)-sulphite system depends on the molar ratio of reagents and the applied light. It was observed that the use of 10-fold excess of Na_2SO_3 in ratio to $Fe_2(SO_4)_3$ and irradiation by solar light resulted in total DOX decomposition in 90 minutes. In order to recognize the main oxidizing agent in the light- Fe(III)-sulphite system the scavenging experiments for the degradation of DOX were performed by adding tert-butyl alcohol (TBA) to the reaction medium. The applied final concentration of TBA was 0.5 mol dm^{-3} while ferric sulphate and sodium sulphite were 10^{-3}, respectively. The kinetic graphs of DOX concentration changes without the presence of TBA outlined in Figure 3, show that DOX degradation by the Fe(III)-sulphite process is mainly caused by sulphate radicals' oxidative action. At the presence of tert-butyl alcohol, the lesser extent of DOX decay was achieved by approximately 20%. This effect was particularly pronounced in an initial stage of the reaction. As the rate of TBA reaction with hydroxyl radicals is approximately

1000-fold greater than that with sulphate radicals [39], it could be concluded that $SO_4^{\cdot-}$ radicals are the major reactive species responsible for DOX degradation.

Table 3. Kinetic parameters of DOX degradation in advanced oxidation systems.

Studied Process	Concentration of H_2O_2/mol dm^{-3}	Concentration of Fe^{2+}/mol dm^{-3}	pH	k/min^{-1}	$t_{1/2}$/min	% of Degradation
UV/H$_2$O$_2$	5×10^{-4}			11.6×10^{-3}	59.7	72
	10^{-4}			11.9×10^{-3}	58.5	73
	10^{-2}	-	8	12.10×10^{-3}	57.2	73.5
	5×10^{-2}			12.50×10^{-3}	55.5	74.5
				k/min^{-1}mol^{-1} dm^3		
Classical Fenton reaction	10^{-4}	10^{-4}		52.5	982	15
	2×10^{-4}	10^{-4}		127.6	535	27
	10×10^{-4}	10^{-4}		332.8	200	48
	10^{-4}	2×10^{-4}		51.2	956	12
	10^{-4}	5×10^{-4}		45.0	1351	11
	10^{-4}	10×10^{-4}		5.5	9823	1.5
Photo-Fenton reaction	10^{-4}	10^{-4}		86.6	657	25
	2×10^{-4}	10^{-4}		296.0	244	46
	10×10^{-4}	10^{-4}		3308.0	31	100
	10^{-4}	2×10^{-4}	3.5	71.8	785	21
	10^{-4}	5×10^{-4}		265.0	239	50
	10^{-4}	10×10^{-4}		53.0	1002	17
	Concentration of Fe$_2$(SO$_4$)$_3$/mol dm^{-3}	Concentration of Na$_2$SO$_3$/mol dm^{-3}				
UV/Fe(III)-SO$_3^{2-}$	5×10^{-5}	10^{-3}		2538	22	61
	5×10^{-5}	2×10^{-3}		1324	41	63
	5×10^{-5}	3×10^{-3}		715	77	59
	5×10^{-5}	4×10^{-3}		394	58	58
	10^{-4}	10^{-3}		7892	7	75
	1.5×10^{-4}	10^{-3}		587	93	66
	2×10^{-4}	10^{-3}		544	100	64
Vis/Fe(III)-SO$_3^2$	10^{-4}	10^{-3}		1986	3	100
Vis/Fe(III)-SO$_3^2$/TBA	10^{-4}	10^{-3}				75

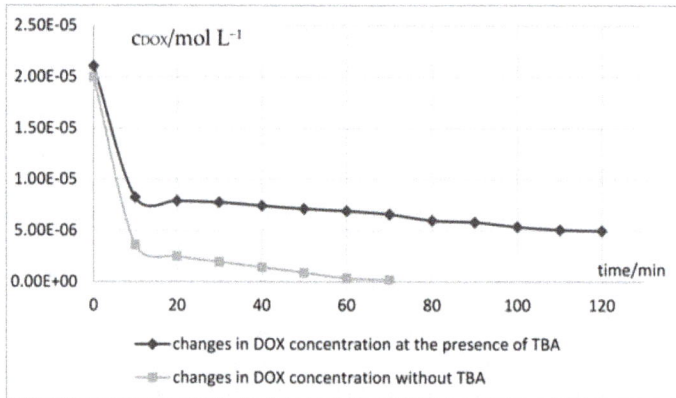

Figure 3. Changes in DOX concentration under influence of Vis-Fe(III)-sulphite system in presence and without TBA.

3.5. A DFT Mechanistic Study of the DOX Decomposition

In order to gain a deeper understanding of the initial stages of the process of molecular degradation of doxazosin, a quantitative DFT mechanistic study was carried out. The calculations were performed with the GAUSSIAN 09 program [40]. To accurately capture properties of the transition states, a DFT functional M06-2X [41], designed especially for the chemical kinetics, was utilized in combination with a Dunning's correlation consistent basis set cc-pVDZ. This theory level is sufficient to capture the physical change along the reaction coordinate for processes investigated here. To model the solvent (water) effect, the CPCM approach was used [42], as implemented in the G09 program. The reaction course is outlined in Figure 4, and all the species involved are pictured in Figure 5. As seen from Figure 4, the reaction is initialized by the addition of the –OH group to the aromatic carbon (atom 7). It is well known that such processes go through short-lived intermediate (thermodynamically controlled step) and transition states (kinetic control) to form an adduct vulnerable to further degradation. It starts with breaking the bond between the addition center and neighboring nitrogen (N10), which is followed by the internal H transfer to form an intermediate (IM3) with keto and amino groups on the bond breaking sites (atoms C7 and N10). The actual destruction of the molecular structure occurs via breaking bond between N3 and C7. Recently, a scheme of the DOX decomposition based on the B3LYP results was proposed [43]. The favorable pathway takes place by cracking bonds N12-C15 and N12-C16. We tried to reproduce this scheme; however, at the M06-2X/cc-pVDZ theory level, no proper transition states leading to ring degradation were found despite many attempts. Nevertheless, since the molecule break-up proposed here starts at the closest vicinity of the bonds N12-C15 and N12-C16 (see Figure 5), our results led to the products with similar molecular masses as those proposed in [43]. As such, their experimental analysis supports both proposed pathways.

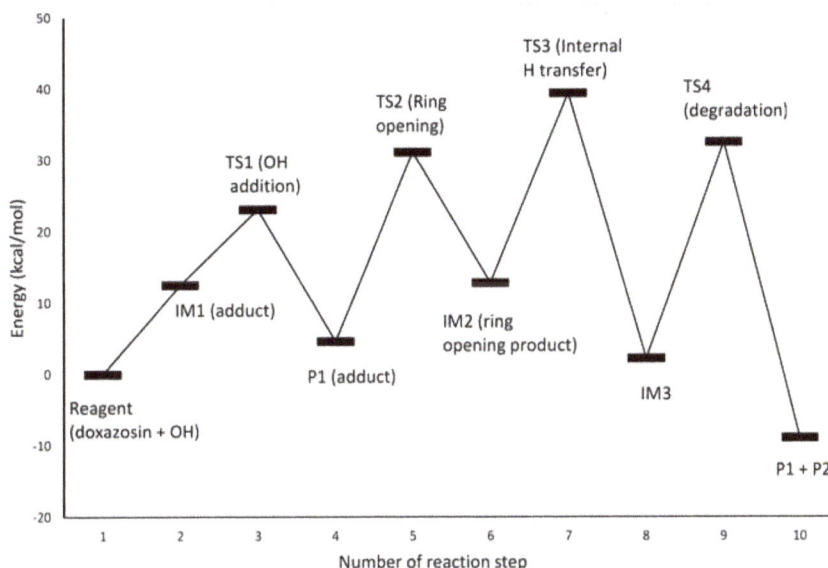

Figure 4. Energy profile of the proposed reaction mechanism (in terms of ΔG values, kcal/mol).

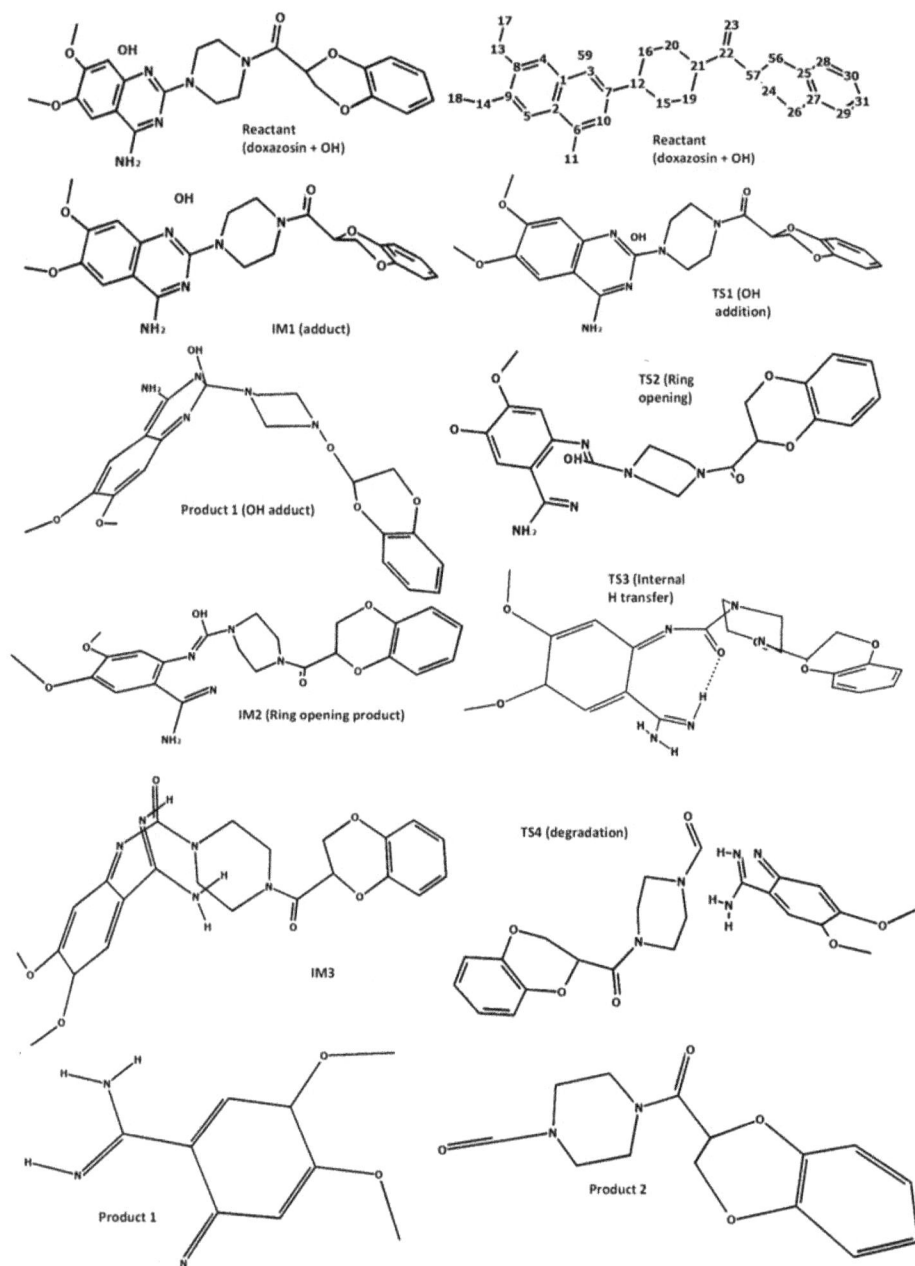

Figure 5. The optimized (at the M06-2X/cc-pVDZ level) geometries of the species involved in the reaction scheme from Figure 4.

4. Conclusions

The obtained results show that doxazosin is a photoliable compound. Experiments done with laboratory solutions demonstrated that DOX direct photolysis is promoted by the basic medium and proceeds faster under the influence of UV light. It was stated that observed degradation of DOX is the result of direct photolysis. The presence of natural matrix acted as a photosensitizer and accelerated the degradation process. The presented surface waters natural radicals made DOX more sensitive to visible light. The run of DOX degradation under influence of some AOP-s were examined. Among the processes taken under consideration, photo-Fenton reaction and Vis/Fe(III)-SO_3^{2-} appeared to be the most efficient.

The DFT mechanistic study provided an understanding of the role of OH^- ion in the photolysis process and pointed out which transformations of molecule lead to its decomposition. It was stated that the initial step of the process is the formation of unstable adduct by bonding the −OH group to the C7 carbon of the aromatic ring. The calculation pointed to the dissociation of the bond between N3 and C7, which lead to molecule disintegration, followed by the internal H transfer and formation of the IM3 intermediate.

Author Contributions: Conceptualization, J.K. (Joanna Karpinska) and A.S.; methodology J.K. (Joanna Karpinska) and A.S.; DFT study, A.R.; investigation, J.K. (Jolanta Koldys), and A.S.; writing—J.K. (Joanna Karpinska), A.R.; writing—review and editing, J.K. (Joanna Karpinska) and A.S.; visualization, A.R. and A.S.; supervision, J.K. (Joanna Karpinska).

Funding: This research received no external funding.

Acknowledgments: The authors would like to thank the Computational Center of the University of Bialystok (Grant GO-008) for providing access to the supercomputer resources and the GAUSSIAN 09 program.

Conflicts of Interest: The authors declare no conflict of interest.

References

1. Generation and potential engineering solutions. *Environ. Technol. Innov.* **2017**, *8*, 40–56. [CrossRef]
2. Dimpe, K.M.; Nomngongo, P.N. Current sample preparation methodologies for analysis of emerging pollutants in different environmental matrices. *Trends Anal. Chem.* **2016**, *82*, 199–207. [CrossRef]
3. Sajid, M.; Płotka-Wasylka, J. Combined extraction and microextraction techniques: Recent trends and future perspectives. *Trends Anal. Chem.* **2018**, *103*, 74–86. [CrossRef]
4. Wang, X.; Huang, P.; Ma, X.; Du, X.; Lu, X. Enhanced in-out-tube solid-phase microextraction by molecularly imprinted polymers-coated capillary followed by HPLC for endocrine disrupting chemical analysis. *Talanta* **2019**, *194*, 7–13. [CrossRef] [PubMed]
5. Aznar, R.; Albero, B.; Sanchez-Brunete, C.; Miguel, E.; Martin-Girela, I.; Tadeo, J.L. Simultaneous determination of multiclass emerging contaminants in aquatic plants by ultrasound-assisted matrix solid-phase dispersion and GC-MS. *Environ. Sci. Pollut. Res.* **2017**, *24*, 7911–7920. [CrossRef]
6. Loos RTavazzi, S.; Mariani, G.; Suurkuusk, G.; Paracchini, B.; Umlauf, G. Analysis of emerging contaminants in water, fish and suspended particulate matter (SPM) in the joint danube survey using solid phase extraction followed by UHPLC-MS-MS and GC analysis. *Sci. Total Environ.* **2017**, *607–608*, 1201–1212. [CrossRef] [PubMed]
7. Jauković, Z.D.; Grujić, S.D.; Bujagić, I.V.M.; Lauśević, M.D. Determination of sterols and steroid hormones in surface water and wastewater using liquid chromatography-atmospheric pressure chemical ionization-mass spectrometry. *Microchem. J.* **2017**, *135*, 39–47. [CrossRef]
8. Ebele, A.J.; Abdallah, M.A.-E.; Harrad, S. Pharmaceuticals and personal care products (PPCPs) in the freshwater aquatic environment. *Emerg. Contracept.* **2017**, *3*, 1–16. [CrossRef]
9. Tran, N.H.; Reinhard, M.; Gin, K.Y.-H. Occurrence and fate of emerging contaminants in municipal wastewater treatment plants from different geographical regions—A review. *Water Res.* **2018**, *133*, 182–207. [CrossRef] [PubMed]
10. Gore, A.C.; Chappell, V.A.; Fenton, S.E.; Flaws, J.A.; Nadal, A.; Prins, G.S.; Toppari, J.; Zoeller, R.T. EDC-2: The endocrine society's second scientific statement on endocrine-disrupting chemicals. *Endocr. Rev.* **2015**, *36*, E1–E150. [PubMed]
11. Blum, K.M.; Andersson, P.L.; Ahrens, L.; Wiberg, K.; Haglund, P. Persistence, mobility and bioavailability of emerging organic contaminants discharged from sewage treatment plants. *Sci. Total Environ.* **2018**, *612*, 1532–1542. [CrossRef]

12. Grandclement, C.; Seyssiecq, I.; Piram, A.; Wong-Wah-Chung, P.; Vanot, G.; Tiliacos, N.; Roche, N.; Doumeng, P. From the conventional biological wastewater treatment to hybrid process, the evaluation of organic micropollutant removal: A review. *Water Res.* **2017**, *111*, 297–317. [CrossRef] [PubMed]

13. Boczkaj, G.; Fernandes, A. Wastewater treatment by means of advanced process at basic pH conditions: A review. *Chem. Eng. J.* **2017**, *320*, 608–633. [CrossRef]

14. Gmurek, M.; Olak-Kucharczyk, M.; Ledakowicz, S. Photochemical decomposition of endocrine disrupting compounds—A review. *Chem Eng. J.* **2017**, *310*, 437–456. [CrossRef]

15. Ismail, L.; Ferronato, C.; Fine, L.; Jaber, F.; Chovelon, J.M. Effect of water constituents on the degradation of sulfaclozine in the three systems UV/TiO$_2$, UV/K$_2$S$_2$O$_8$, and UV/TiO$_2$/K$_2$S$_2$O$_8$. *Environ. Sci. Pollut. Res.* **2018**, *25*, 2561–2663. [CrossRef] [PubMed]

16. Fast, S.A.; Gude, V.G.; Truax, D.D.; Martin, J.; Magbanua, B.S. A critical evaluation of advanced oxidation processes for emerging contaminants removal. *Environ. Process* **2017**, *4*, 283–302. [CrossRef]

17. Omar, M.A.; Mohamed, A.-M.I.; Derayea, S.M.; Hammad, M.A.; Mohamed, A.A. An efficient spectrofluorimetric method adopts doxazosin, terazosin and afluzosin coupling with orthophtalaldehyde: Application in human plasma. *Spectochim. Acta Part A* **2018**, *195*, 215–222. [CrossRef] [PubMed]

18. Jekell, A.; Kalani, M.; Kahan, T. The effects of alpha 1-adrenoceptor blockade and angiotensyn converting enzyme inhibition on central and brachial blood pressure and vascular reactivity: The doxazosin-ramipril study. *Heart Vessels* **2017**, *32*, 674–684. [CrossRef]

19. Anzenbacher, P.; Zanger, U.M. *Metabolism of Drugs and Other Xenobitics*; John Willey & Sons: Weinheim, Germany, 2012.

20. Ojha, T.; Bakshi, M.; Charkraborti, A.K.; Singh, S. The ICH guidance in practice: Stress decomposition studies on three piperazinyl quinazoline adrenergic receptor—Blocking agents and comparison of their degradation behaviour. *J. Pharm. Biomed Anal.* **2003**, *31*, 775–783. [CrossRef]

21. Zafirou, O.C.; Joussot-Dubien, J.; Zepp, R.G.; Zika, R.G. Photochemistry of natural waters. *Environ. Sci. Technol.* **1984**, *18*, 358A–371A. [CrossRef]

22. Chen, Y.; Liu, L.; Liang, J.; Wu, B.; Zuo, J.; Zuo, Y. Role of humic substances in the photodegradation of naproxen under simulated sunlight. *Chemosphere* **2017**, *187*, 261–267. [CrossRef] [PubMed]

23. Canonica, S. Oxidation of organic contaminants induced by excited triplet states. *Chimia* **2007**, *61*, 641–644. [CrossRef]

24. Yang, Y.; Pignatello, J. Participation of halogens in photochemical reactions in natural and treated waters. *Molecules* **2017**, *22*, 1684. [CrossRef] [PubMed]

25. Zhang, K.; Parker, K.M. Halogen radical oxidants in natural and engineered aquatic systems. *Environ. Sci. Technol.* **2018**, *52*, 9579–9594. [CrossRef] [PubMed]

26. Devi, L.G.; Munikrishnappa, C.; Nagraj, B.; Rajashekhar, K.E. Effect of chloride and sulfate ions on the advanced photo Fenton and modified photo-Fenton degradation of alizarin red S. *J. Mol. Catal. A Chem.* **2013**, *374–375*, 125–131. [CrossRef]

27. Zepp, R.G.; Holgne, J.; Bader, H. Nitrate-induced photooxidation of trace organic chemicals in water. *Environ. Sci. Technol.* **1987**, *21*, 443–450. [CrossRef]

28. Calza, P.; Vione, D.; Novelli, A.; Pelizzetti, E.; Minero, C. The role of nitrite and nitrate ions as photosensitizers in the phototransformation of phenolic compounds in seawater. *Sci. Total Environ.* **2012**, *439*, 67–75. [CrossRef]

29. Machutek, A.J.; Morales, J.E.F.; Okano, L.T.; Silvero, C.A.; Quina, F.H. Photolysis of ferric ions in the presence of sulfate or chloride ions: Implications for the photo-Fenton process. *Photochem. Photobiol. Sci.* **2009**, *8*, 985–991.

30. Zepp, R.G.; Faust, B.C.J.; Holgne, J. Hydroxyl formation in aqueous reactions (pH 3–8) of iron(II) with hydrogen peroxide: The photo-Fenton reaction. *Environ. Sci. Technol.* **1992**, *26*, 313–319. [CrossRef]

31. Gao, H.; Zepp, R.C. Factors influencing photoreactions of dissolved organic matter in a coastal river of the Southeastern United States. *Environ. Sci. Technol.* **1998**, *32*, 2940–2946. [CrossRef]

32. Zepp, R.C.; Schlozhauerm, P.F.; Simmons, M.S.; Miller, G.C.; Baughman, G.L.; Wolfe, N.L. Dynamics of pollutant photoreactions in hydrosphere. *Fresenius Z. Anal. Chem.* **1984**, *319*, 119–125. [CrossRef]

33. Legrini, O.; Oliveros, E.; Braun, A.M. Photochemical processes for water treatment. *Chem. Rev.* **1993**, *93*, 671–698. [CrossRef]

34. Jain, B.; Singh, A.K.; Kim, H.; Lichtfouse, E.; Sharma, V.K. Treatment of organic pollutants by homogenous and heterogeneous Fenton reaction process. *Environ. Chem. Lett.* **2018**, *16*, 947–967. [CrossRef]

35. Melin, V.; Hentiquez, A.; Radojkovic, C.; Schwederski, B.; Kaim, W.; Frerr, J.; Contreras, D. Reduction reactivity of catecholamines and their ability to promote a Fenton reaction. *Inorg. Chim. Acta* **2016**, *453*, 1–7. [CrossRef]

36. Wang, J.; Wang, S. Activation of persulfate (PS) and peroxymonosulfate (PMS) and application for the degradation of emerging contamination. *Chem. Eng. J.* **2018**, *334*, 1502–1517. [CrossRef]

37. Devi, P.; Das, U.; Dalai, A.K. In-situ chemical oxidation: Principle and applications of peroxide and persulfate treatments in wastewater systems. *Sci. Total Environ.* **2016**, *571*, 643–657. [CrossRef] [PubMed]

38. Matzek, L.W.; Carter, K.E. Activated persulfate for organic chemical degradation: A review. *Chemosphere* **2016**, *151*, 178–188. [CrossRef]

39. Guo, Y.; Lou, X.; Fang, C.; Xiao, D.; Wang, Z.; Liu, J. Novel photo-sulfite system: Toward simultaneous transformations of inorganic and organic pollutants. *Environ. Sci. Technol.* **2013**, *47*, 11174–11181. [CrossRef]

40. Frisch, M.J.; Trucks, G.W.; Schlegel, H.B.; Scuseria, G.E.; Robb, M.A.; Cheeseman, J.R.; Scalmani, G.; Barone, V.; Mennucci, B.; Petersson, G.A.; et al. *Gaussian 09, Revision D*; Gaussian, Inc.: Wallingford, CT, USA, 2009.

41. Zhao, Y.Y.; Truhlar, D.G. The M06 suite of density functionals for main group thermochemistry, thermochemical kinetics, noncovalent interactions, excited states, and transition elements: Two new functionals and systematic testing of four M06-class functionals and 12 other functionals. *Theor. Chem. Acc.* **2008**, *120*, 215–241.

42. Barone, W.; Cossi, M.; Tomasi, J. Geometry optimization of molecular structures in solution by the polarizable continuum model. *J. Comput. Chem.* **1998**, *19*, 404–417. [CrossRef]

43. El-Desawy, M.; Zayed, M.A.; Ferrag, J. Fragmentation pathway of doxazosin drug: Thermal analysis, mass spectrometry and DFT calculations and NBO analysis. *J. Pharm. Appl. Chem.* **2017**, *3*, 45–51. [CrossRef]

water

MDPI

Article

Pristine and Graphene-Quantum-Dots-Decorated Spinel Nickel Aluminate for Water Remediation from Dyes and Toxic Pollutants

Elzbieta Regulska *[ID], Joanna Breczko and Anna Basa[ID]

Institute of Chemistry, University of Bialystok, Ciolkowskiego 1K, 15-245 Bialystok, Poland;
j.luszczyn@uwb.edu.pl (J.B.); abasa@uwb.edu.pl (A.B.)
* Correspondence: e.regulska@uwb.edu.pl

Received: 31 March 2019; Accepted: 1 May 2019; Published: 7 May 2019

check for
updates

Abstract: Pristine nickel aluminate and the one decorated with graphene quantum dots were prepared via a cost-effective co-precipitation method. Both were fully characterized by thermogravimetry (TGA), differential scanning calorimetry (DSC), attenuated total reflectance Fourier transform infrared (ATR-FTIR) spectroscopy, X-ray diffraction (XRD), scanning electron microscopy (SEM), energy-dispersive X-ray (EDX) spectroscopy, transmission electron microscopy (TEM), and UV–Vis techniques. The photocatalytic activity of nickel aluminate under simulated solar light irradiation was demonstrated towards potential pollutants, including a series of dyes (rhodamine B, quinoline yellow, eriochrome black T, methylene blue), toxic phenol and fungicide (thiram). Further profound enhancement of the photocatalytic activity of nickel aluminate was achieved after its decoration with graphene quantum dots. The mechanism of the photocatalytic degradation in the presence of the $NiAl_2O_4$/graphene quantum dots (GQDs) composite was investigated; hydroxyl radicals were found to play the leading role. This work offers new insight into the application of the conjunction of the inorganic spinel and the carbon nanostructure (i.e., GQDs), but also provides a simple and highly efficient route for potential water remediation from common pollutants, including dyes and colorless harmful substances.

Keywords: graphene quantum dots; nickel aluminate; photocatalysis; spinel; water remediation

---

1. Introduction

Nanocrystalline spinel aluminates with the general formula of MAl_2O_4 (M = Ni, Zn, Mn, Co, Mg, etc.) attract research interest due to their versatile properties. Aluminates have high thermal stability, mechanical resistance, hydrophobicity and low surface acidity. Nickel aluminates are one of the most important aluminate materials, and have been studied for their many applications, including electrochemical sensing [1], pigments [2], catalysts [3–7], photocatalysts [8–12], magnetic [13] and refractory materials [14]. $NiAl_2O_4$ has also attracted attention as a hydrogen storage material [15,16], oxygen carrier in combustion loop reactors [17] and as a component of supercapacitor electrode materials [18]. Very recently, attention has been focused on its photocatalytic performance [10–12]. First reports described the utilization of nickel-aluminum layered hydroxides for carbon dioxide conversion [19,20] and for dye degradation [21,22]. However, to the best of our knowledge the photocatalytic activity of the nickel aluminate of the spinel structure was demonstrated for the first time in 2015 by A. Sobhani-Nasab et al. [8]. The catalyst was synthesized via sol-gel method, to then be applied for methyl orange degradation under visible light irradiation. M. Rahimi-Nasrabadi et al. [9] performed analogous experiments using ultraviolet light. The photocatalytic activity of the nickel aluminate spinel against a series of dyes (i.e., rhodamine B (RhB), methylene blue (MB), methyl orange (MO), methyl red) was demonstrated by T. Tangcharoen et al. [10]. In the past year, the enhanced photocatalytic performance of NiAl$_2$O$_4$ was achieved by its substitution with copper, nickel and magnesium ions [11,12]. These photocatalytic tests were run against MB, MO and Congo red, both using ultraviolet and solar light irradiation.

In our studies we decided to utilize carbon nanostructures, namely graphene quantum dots (GQDs), to decorate spinel nickel aluminate in order to achieve enhanced photocatalytic performance of the composite. Carbon nanoforms, including fullerenes [23–26], carbon nanotubes [27], graphite [28], graphene [29], graphene oxide [30] and graphene quantum dots [31], have already been used for this purpose. GQDs are nanoscale (diameter < 100 nm) fragments of graphene, revealing size-dependent luminescence. Due to their extended π-electron system, GQDs broaden the spectral range that can be harvested by the composite. Additionally, thanks to discrete electronic levels, they allow for hot electron injection and efficient charge separation. Moreover, GQDs can be prepared from cheap and accessible precursors [31,32]. Therefore, GQDs have already been used to form composites with inorganic semiconductors, which exhibited photocatalytic activity [33,34]. Nevertheless, to the best of our knowledge nickel and aluminum oxides or mixed Ni/Al oxides have not been examined in that context.

On the other hand, the constantly rising production of municipal and industrial wastes and a constant struggle with ineffective methods of water remediation requires us to seek new efficient and economically attractive procedures of wastewater treatment. Heterogenous photocatalysis utilizing solid catalysts that are active under solar light irradiation provides hope that it is possible to meet all mentioned requirements. Therefore, we decided to combine the attractive properties of spinel nickel aluminate and graphene quantum dots by preparing a NiAl$_2$O$_4$/GQDs composite for superior photocatalytic performance. Our report aimed to achieve several goals: (i) reveal the universality of nickel aluminate as a photocatalyst; reveal its capability to degrade pollutants, forming both colorful and colorless aqueous solutions; (ii) enhance its photocatalytic activity by decorating it with graphene quantum dots (GQDs); (iii) assess photocatalyst activity under solar light irradiation; (iv) examine the photodegradation mechanism in the presence of the NiAl$_2$O$_4$/GQDs composite.

2. Materials and Methods

2.1. Materials

Ammonium oxalate, aluminum nitrate nonahydrate, dimethyl sulfoxide, isopropyl alcohol, nickel(II) nitrate hexahydrate, phenol, quinoline yellow, RhB, terephthalic acid and thiram were purchased from Sigma-Aldrich (Warsaw, Poland). MB was supplied by Park Scientific (Northampton, UK). Aqueous ammonia, citric acid monohydrate, eriochrome black and sodium hydroxide were obtained from POCh (Gliwice, Poland).

2.2. Synthetic Procedures

2.2.1. Synthesis of NiAl$_2$O$_4$

Nickel aluminate catalyst was prepared by a coprecipitation method. Nickel and aluminum were coprecipitated from an aqueous nitrate solution (5 mmol·L^{-1}) by adding an aqueous ammonia (1 mol·L^{-1}) solution to produce a pH of about 8. The molar ratio of Al/Ni in the solution was 2.0. The precipitates were collected by filtration, washed with water and dried at 100 °C for 6 h. The as-synthesized nickel aluminate was subjected to calcination in air at 400, 600 and 800 °C for 3 h to obtain the crystalline spinel.

2.2.2. Synthesis of GQDs and NiAl$_2$O$_4$/GQDs

GQDs were prepared according to the procedure described elsewhere [32]. The NiAl$_2$O$_4$/GQDs composite was formed by introducing the crystalline NiAl$_2$O$_4$ spinel during the synthesis of GQDs.

Briefly, 0.62 mmol of citric acid monohydrate and 0.85 mmol of the crystalline spinel were heated to 200 °C for 30 min until the transparent liquid changed color through yellow to amber. Subsequently, the heating temperature was reduced to 140 °C, and 10 mL of deionized water were added. The obtained mixture was heated under stirring until the complete evaporation of water.

2.3. Methods

DSC and TGA analyses were performed by Thermal Analyzer TGA/DSC 1 (METTLER TOLEDO, Giessen, Germany) with a heating rate of 15 °C·min^{-1} under a nitrogen environment with a flow rate of 20 mL·min^{-1}. All runs were carried out from 25 to 1550 °C. The measurements were made in alumina crucibles with lids.

The powder X-ray diffraction data were measured at 293 K using a SuperNova diffractometer (Rigaku, The Woodlands, TX, USA) with a charge-coupled device (CCD) and a Cu-Kα radiation source at a150 mm sample-to-detector distance.

Scanning electron microscopy images were recorded by secondary-electron SEM with the use of an INSPECT S50 scanning electron microscope (FEI, Hillsboro, OR, USA). The accelerating voltage of the electron beam was 15 keV and the working distance was 10 mm. Images were also obtained with a TEM system (FEI Teknai T20 G2 X-TWIN, Hillsboro, OR, USA) operating at 200 kV, equipped with an LaB$_6$ source.

The ATR-FTIR spectra (3200–500 cm^{-1}) were obtained using a Nicolet Model 6700 FT-IR spectrometer with a DTGS detector (Thermo Scientific, Madison, WI, USA). The crystal-diamond spectra were obtained with 4 cm^{-1} resolution, and 32 scans for each sample spectrum were obtained. Diffuse reflectance UV–Vis spectra (DRS) were recorded on a Jasco V-30 UV–Vis/NIR spectrophotometer (Jasco, De Meern, Netherlands) equipped with an integrating sphere 60 mm in diameter using BaSO$_4$ as a reference.

The UV–Vis spectra were recorded with a HITACHI U-2800A UV–Vis spectrophotometer (Hitachi, Tokyo, Japan) equipped with a double monochromator and a single-beam optical system (190–700 nm). A SUNTEST CPS+ (ATLAS, Mount Prospect, IL, USA) solar simulator apparatus was used to perform photocatalytic degradation experiments. The emission spectra were recorded on a Hitachi F-7000 fluorescence spectrophotometer (Hitachi, Tokyo, Japan): excitation bandwidths 5.0 nm; emission bandwidths 5.0 nm; scan speed 1200 nm·min^{-1}.

2.4. Photocatalysis Experiments

The photocatalytic degradation experiments were investigated in a 50-mL glass cell. The reaction mixture consisted of 20 mL of the model pollutant RhB (2×10^{-5} mol·L^{-1}), quinoline yellow (QY, 2×10^{-5} mol·L^{-1}), eriochrome black (EB, 4×10^{-5} mol·L^{-1}), MB (8×10^{-5} mol·L^{-1}), phenol (PH, 10^{-4} mol·L^{-1}) or thiram (TM, 8×10^{-5} mol·L^{-1}), aqueous solution and a photocatalyst (1.5 g·L^{-1}). Prior to photocatalytic experiments, the catalyst was settled in suspension for 60 min in the dark for the adsorption–desorption equilibrium. All the above-mentioned chemicals were analytical-grade reagents and used without further treatment. All solutions were prepared using deionized water, which was obtained by a Polwater apparatus (Polwater, Cracow, Poland).

2.5. Terephthalic Acid Probe Method

The generation of hydroxyl radicals as a consequence of the irradiation of the aqueous suspension of NiAl$_2$O$_4$ catalyst with the simulated solar light was examined using the terephthalic acid (TPA) probe method [35]. The nickel aluminate particles (2 mg·mL^{-1}) were dispersed in a 3 mmol·L^{-1} TPA solution prepared in a 10 mmol·L^{-1} NaOH solution. Afterwards, the obtained suspension was sonicated for 10 min and exposed to sunlight for 2 h while vigorous stirring continued. After a given time the suspension was centrifuged, and the fluorescence emission spectrum was measured at the excitation wavelength of $\lambda = 312$ nm.

2.6. Reactive Species Scavenging

The generation of electron holes, hydroxyl radicals and electrons was determined by treating the reaction suspensions with ammonium oxalate (AO), isopropyl alcohol (IPA) and dimethyl sulfoxide (DMSO) as respective scavengers [36].

3. Results and Discussion

3.1. Structural and Morphological Study

3.1.1. NiAl$_2$O$_4$

Thermogravimetric studies were performed to examine the temperature required for the formation of the crystalline form of nickel aluminate. Figure 1 presents DSC, TGA and derivative thermogravimetry (DTG) curves of the as-synthesized NiAl$_2$O$_4$ before annealing. The TGA curve shows distinct mass loss in the temperature range of 230–400 °C, represented by the peak on the DTG curve at 284 °C. In the same temperature window, exothermic peaks on the DSC curve were attributed to the decomposition of the Ni(Ac)$_2$·4H$_2$O and the following structural ordering of the nickel aluminate spinel phase. Therefore, it was concluded that 400 °C should be the lowest temperature used for the annealing to obtain the stable final inorganic crystalline product. Accordingly, the as-synthesized nickel aluminate was divided into three portions, which were annealed respectively at 400, 600 and 800 °C.

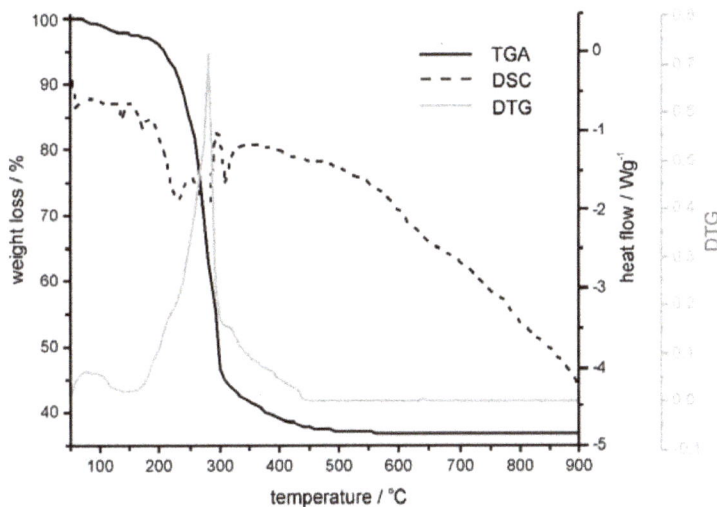

Figure 1. Differential scanning calorimetric (DSC), thermogravimetric (TGA) and derivative thermogravimetric (DTG) curves of the as-synthesized NiAl$_2$O$_4$ before annealing.

XRD images of the NiAl$_2$O$_4$ annealed at 400, 600 and 800 °C (Figure 2) showed that the contribution of the spinel structure increased with increasing applied temperature. Nickel oxide was observed at lower temperatures (400 and 600 °C), as indicated by diffraction patterns assigned to [200] and [220] lattices (JCPDS No. 47-1049). The pure spinel crystalline form of nickel aluminate (well-matched with JCPDS No. 44-0160), with no remaining cubic NiO, was obtained after annealing at 800 °C. The apparent crystallite diameter (D_c) of NiAl$_2$O$_4$ particles was found to be of 3, 4 and 8 nm for samples annealed at 400, 600 and 800 °C, respectively. The latter numbers were calculated according to the Scherrer equation: $D_c = k\lambda/\beta\cos\theta$, where β is the full width at the half maximum of the diffraction peak, k is the empirical constant (0.9), θ is the angular position of the diffraction peak, and λ is the wavelength of the X-ray source (here 1.5405 Å). The observed increase of crystallite sizes with the increase of the annealing temperature was in agreement with the findings of others reported for inorganic semiconductors (e.g., nickel oxide) [37].

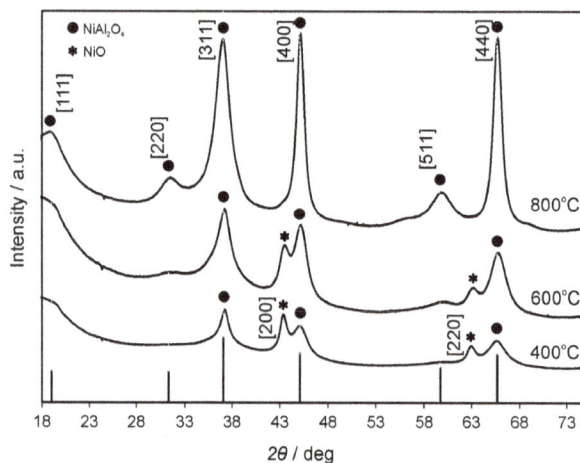

Figure 2. X-ray diffraction (XRD) patterns of NiAl$_2$O$_4$ annealed at 400, 600 and 800 °C.

ATR-FTIR spectroscopy studies were undertaken to confirm the purity of nickel aluminate nanoparticles and to investigate the presence of the functional groups on their surface. ATR-FTIR spectra of the samples annealed at 400, 600 and 800 °C were registered in the range of 500–3600 cm^{-1} (Figure 3). The bands observed at low frequencies within 500–700 cm^{-1} were attributed to the stretching vibrations of Ni–O, Al–O and Ni–O–Al bonds [8,38]. Moreover, the bands observed in the range of 3200–3500 cm^{-1} indicated the presence of the O–H surface bonds on the catalyst surface. The bands depicted in Figure 3 were observed for the samples annealed at all applied temperatures, from 400 to 800 °C. However, their intensity increased alongside the increase of the applied annealing temperature, indicating well-developed crystalline structures for nickel aluminate annealed at 800 °C.

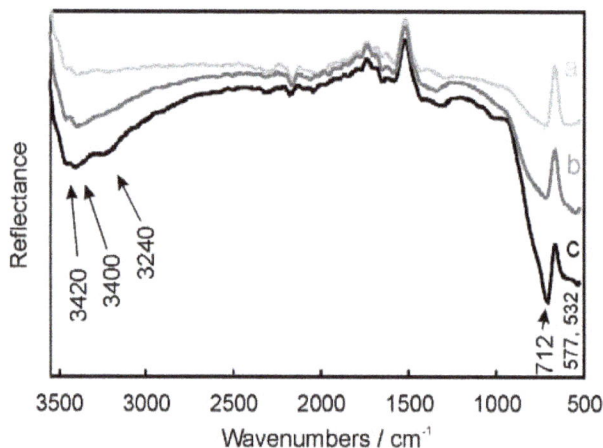

Figure 3. Attenuated total reflectance Fourier transform infrared (ATR-FTIR) spectra of NiAl$_2$O$_4$ after annealing at (**a**) 400, (**b**) 600 and (**c**) 800 °C.

The morphology of the NiAl$_2$O$_4$ particles was evaluated based on SEM analysis. Figure 4 demonstrates that as-synthesized nickel aluminate formed agglomerates with particle sizes in the range of 50–200 μm. However, after annealing the size of the agglomerates decreased with increasing temperature. The highest homogeneity was observed for the material annealed at 800 °C (Figure 4(4a,b)).

The chemical composition and purity of the synthesized nickel aluminate were evaluated using EDX analysis. As shown in Figure 4(1c–4c), Ni, O and Al were the only observed elements in all of the registered curves. Moreover, the decrease of the intensity of the peak attributed to the O element was noticed after annealing. This was due to the formation of the crystalline form of the spinel structure.

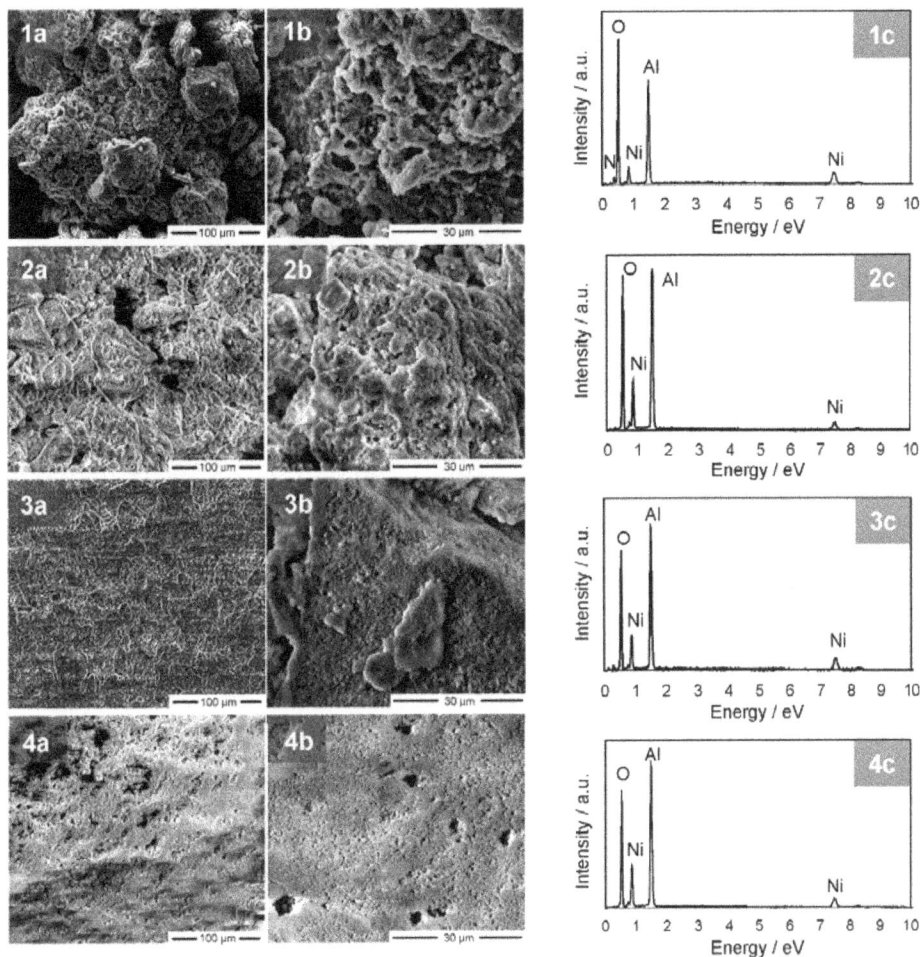

Figure 4. Scanning electron microscopy (SEM) images and X-ray (EDX) spectra of NiAl$_2$O$_4$ without (**1a–c**) and after annealing at (**2a–c**) 400, (**3a–c**) 600 and (**4a–c**) 800 °C.

3.1.2. GQDs and NiAl$_2$O$_4$/GQDs Composite

The pristine GQDs exhibited structure of high porosity, as shown in SEM images (Figure 5(1a,b)), as distinguished from the crystalline NiAl$_2$O$_4$ (Figure 4(4a,b)). Therefore, an increase in the porosity of the NiAl$_2$O$_4$/GQDs composite compared to the pristine spinel was observed, as shown in Figure 5(2a,b). TEM images of GQDs particles (Figure 5(1c,d)) demonstrated their uniform sizes ranging from 2 to

7 nm. However, due to profound differences in the size of GQDs and metal oxide particles, carbon nanostructures could not be distinguished in the TEM image of the NiAl$_2$O$_4$/GQDs. Nevertheless, their presence in the composites led to an increase in the dispersity of the spinel nanoparticles (Figure 5(2c)). The TEM image of the pristine NiAl$_2$O$_4$ (Figure 5(3)) shows an agglomerated structure with particles having an average size of 20 nm. On the other hand, the particles of the NiAl$_2$O$_4$/GQDs (Figure 5(2c)) had smaller diameters (between 7–10 nm) and appeared to be separate from each other. The carbon content in GQDs and in the NiAl$_2$O$_4$/GQDs composites was examined by EDX (see Figure 6). This showed that GQDs presented a moderate oxygen content of 16%. The value of the latter is relevant for photocatalytic activity since surface oxygen groups contribute to the photocatalytic activity on defect sites [39].

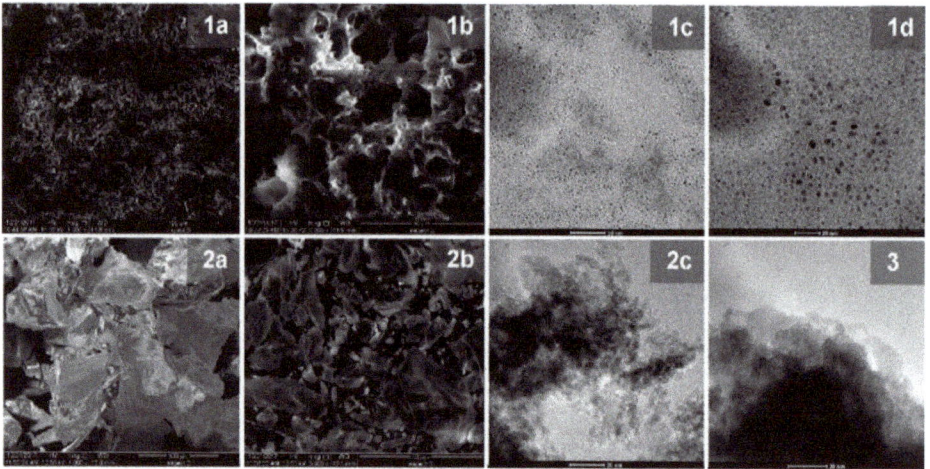

Figure 5. Images of graphene quantum dots (GQDs). (**1a,b**) SEM, (**1c,d**) transmission electron microscopy (TEM); NiAl$_2$O$_4$/GQDs: (**2a,b**) SEM, (**2c**) TEM; and NiAl$_2$O$_4$: (**3**) TEM.

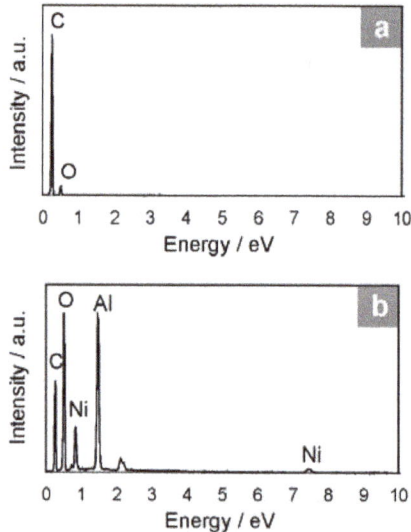

Figure 6. EDX spectra of (**a**) GQDs and (**b**) NiAl$_2$O$_4$/GQDs.

The UV–Vis diffuse reflectance spectrum (Figure 7) of the crystalline $NiAl_2O_4$ (after annealing at 800 °C) displayed a significant absorption in the ultraviolet spectrum range. Additionally, absorption in the visible region, due to the d–d transition of Ni(II) and Al(III) was seen. As typical for the normal spinel structure with the tetrahedrally coordinated Ni(II) in the $NiAl_2O_4$ lattice absorption, a maximum around 650 nm was found. However, a presence of the inverse spinel structure was also revealed, as indicated by the absorption appearing around 380 and 770 nm. This is known to arise from the octahedral Ni(II) ions [40]. Based on the extrapolation of the linear part of the Kubelka–Munk vs. energy plot, the energy bandgap was calculated to be 2.9 and 2.5 eV for $NiAl_2O_4$ and the $NiAl_2O_4$/GQDs composite, respectively. The calculated E_g value of the pristine nickel aluminate was close to that reported for spinel (see Table 1). Meanwhile, the synthesized $NiAl_2O_4$/GQDs composite showed a significantly narrower band edge, which corresponded to 470 nm. This wavelength was in the solar spectrum range of the highest intensity [41], indicating a significant potential to harvest renewable solar energy.

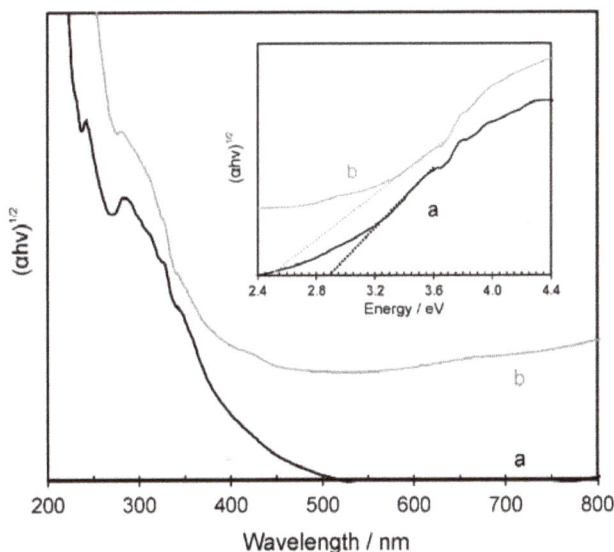

Figure 7. UV–Vis diffuse reflectance profile of the (a) $NiAl_2O_4$ and the (b) $NiAl_2O_4$/GQDs composite annealed at 800 °C.

Table 1. Energy band gap (*Eg*) of the $NiAl_2O_4$/GQDs composite and $NiAl_2O_4$ reported in this work and elsewhere.

Catalyst	$NiAl_2O_4$						$NiAl_2O_4$/GQDs
Ref.	[8]	[11]	[9]	[10]	[13]	This work	This work
Eg/eV	2.85	3.0	3.1	3.45	3.41	2.9	2.5

3.2. Photocatalytic Activity Study

The photocatalytic activity of the $NiAl_2O_4$ nanoparticles was tested against a series of potential water pollutants. These included a series of dyes (i.e., RhB, QY, EB and MB), along with PH and the commonly used fungicide TM. The degradation efficiency is illustrated in Figure 8A as a decrease of the residual concentration ratio (C_t/C_0) of each compound during the time of irradiation with the simulated solar light. All examined model contaminants were found to decompose under the applied conditions. The degradation of all model pollutants followed pseudo–first-order kinetics. Therefore, based on the plots presented in Figure 8B, the pseudo-first-order rate constants were calculated and are compared in Table 2. The determined k values increased in the following order: RhB < QY < EB < PH < TM < MB. Among dyes, the most resistant turned out to be RhB, while MB decomposed the easiest. The resistance to photocatalytic decomposition of MB was close that of TM. Tetramethylthiuram disulfide, unlike PH—which represents aromatic compounds—underwent photo-oxidation easily.

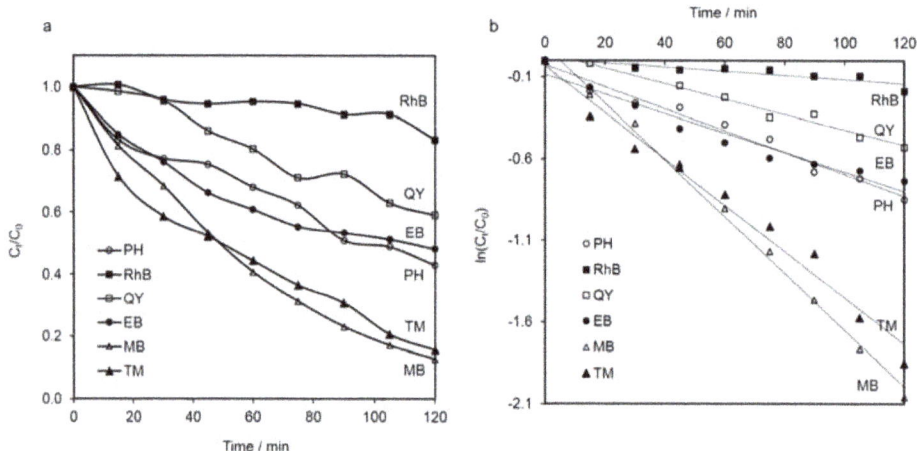

Figure 8. (**a**) Residual concentration ratio (C_t/C_0) and (**b**) apparent first-order kinetic lines of rhodamine B (RhB), quinoline yellow (QY), eriochrome black (EB), methylene blue (MB), phenol (PH) and thiram (TM) as a function of time under simulated solar light irradiation in the presence of $NiAl_2O_4$.

Table 2. The apparent first-order rate constants k (min^{-1}) for the degradation of RhB, QY, EB, MB, PH and TM under simulated solar light irradiation in the presence of $NiAl_2O_4$.

Sample	Rate Constants k/h^{-1}					
	RhB	QY	EB	MB	PH	TM
$NiAl_2O_4$	0.068	0.282	0.354	1.044	0.401	0.852

The photocatalytic activity of the $NiAl_2O_4$/GQDs was examined towards RhB as a representative dye and towards PH (representative of toxic compounds forming colorless aqueous solutions). Each of the chosen model pollutants from the two examined groups exhibited the most resistance to degradation. The results of the photocatalytic studies obtained in the presence of the synthesized composite were compared with those performed using pristine spinel (see Figure 9).

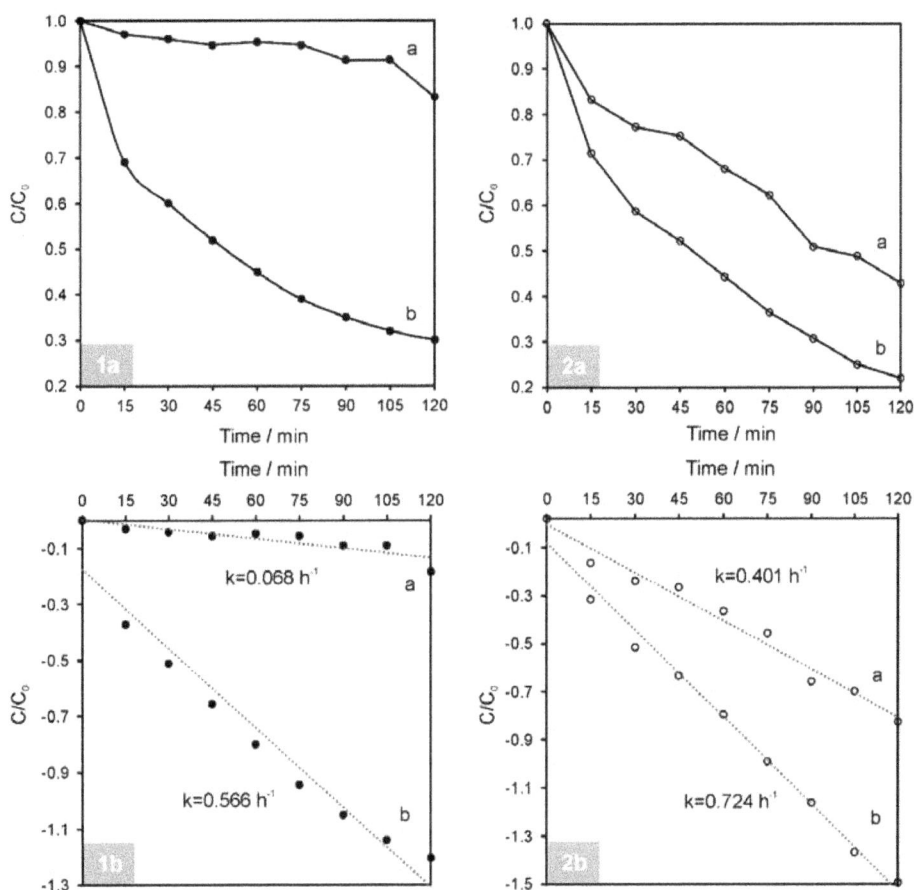

Figure 9. Residual (**1a** and **2a**) concentration ratio (C_t/C_0) and (**1b** and **2b**) kinetic curves as a function of time under the simulated solar light irradiation of (**1**) RhB and (**2**) PH aqueous solutions in the presence of (**a**) NiAl$_2$O$_4$ and (**b**) NiAl$_2$O$_4$/GQDs.

To examine the mechanism of the photocatalytic activity of NiAl$_2$O$_4$/GQDs, a composite hydroxyl radical generation probe method with TPA was applied. Figure 10 shows fluorescence spectra as observed for the supernatant solution of the NiAl$_2$O$_4$/GQDs catalyst suspension irradiated with terephthalate (TP) for various durations. A strong fluorescence emission peak was observed at λ_{em} = 426 nm. This was assigned to the formation of an adduct (hTP) between TP and hydroxyl radical (Scheme 1), indicating the formation of •OH species in the irradiated suspension. The intensity of the observed emission peak increased linearly within the irradiation time, as shown in the inset of Figure 10.

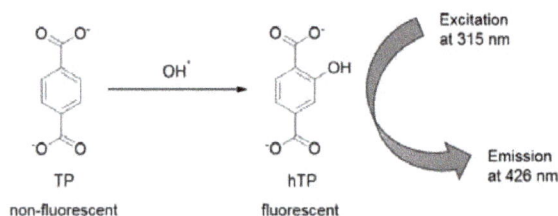

Scheme 1. Formation of fluorescent 2-hydroxyterephthalate (hTP) via the reaction of hydroxyl radicals with terephthalate (TP).

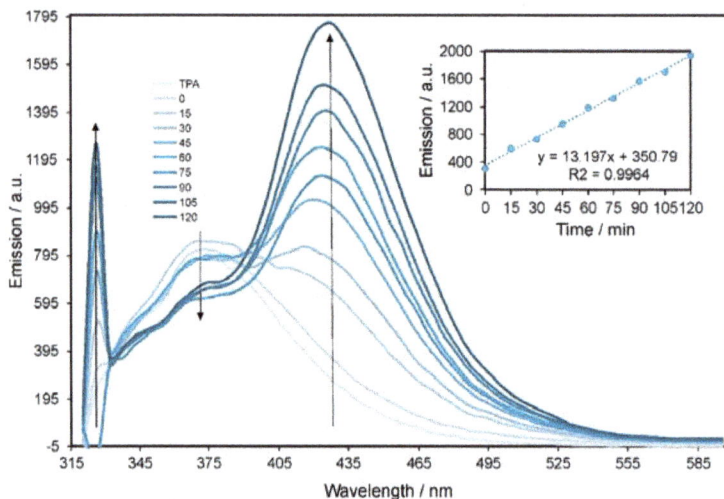

Figure 10. Fluorescence spectra of the solution of terephthalic acid (TPA) under simulated solar light irradiation in the presence of the NiAl$_2$O$_4$/GQDs catalyst within 2 h. Inset: time dependence of the fluorescence intensity at 426 nm.

In order to investigate which other active species were contributing to the photocatalytic activity of NiAl$_2$O$_4$/GQDs, a series of experiments with established scavengers was performed. Ammonium oxalate (AO), isopropyl alcohol (IPA) and dimethylsulfoxide (DMSO) as electron hole, hydroxyl radical and electron scavenger, respectively, were separately mixed with the reactant mixture containing RhB and NiAl$_2$O$_4$. RhB was subjected to photocatalytic degradation under simulated solar light. As shown in Figure 11, the biggest influence on the photocatalytic degradation of RhB was observed in the presence of hydroxyl radicals. However, since AO (being the hole scavenger) also had a significant influence, it indicated that hydroxyl radicals were generated involving both valence band holes and conduction band electrons. The smallest effect was observed in the presence of DMSO, which may point to the instant reaction of the electrons in the conduction band after excitation of the semiconductor. These observations indicated the low electron–hole recombination effect in the synthesized catalyst. The suggested mechanism of the photocatalytic degradation of the organic pollutants in the presence of the NiAl$_2$O$_4$/GQDs composite is presented in Scheme 2. It shows that after GQDs harvest the sunlight, they give rise to the generation of the electron–hole pairs. The same phenomenon occurs in NiAl$_2$O$_4$ since it also absorbs light from the visible spectrum range. Subsequently, the electrons injected in the conduction band of NiAl$_2$O$_4$ may react with oxygen and lead to the generation of hydroxyl radicals, as shown in the Scheme 2. GQDs prolong the recombination rate of the charge carriers. They also contribute to harvesting the sunlight and are responsible for the adsorption of the pollutants, which ultimately decompose.

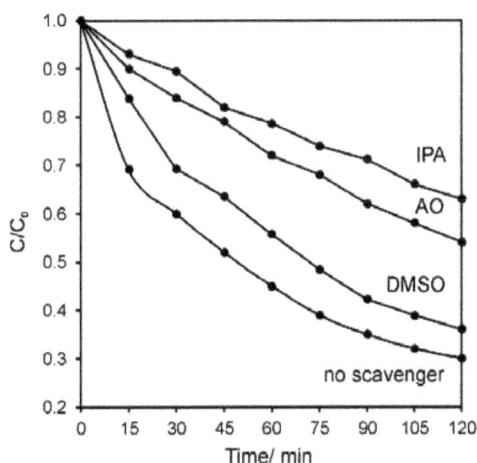

Figure 11. Photocatalytic degradation of RhB in the presence of the NiAl$_2$O$_4$/GQDs composite under simulated solar light irradiation without and in the presence of scavengers: ammonium oxalate (AO), isopropyl alcohol (IPA) and dimethyl sulfoxide (DMSO) used to capture holes, hydroxyl radicals and electrons, respectively.

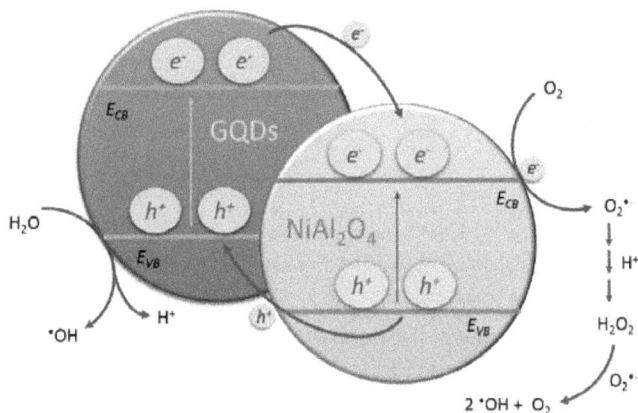

Scheme 2. The proposed mechanism of photocatalysis using the NiAl$_2$O$_4$/GQDs composite.

The reusability of NiAl$_2$O$_4$/GQDs was studied in four successive recycling experiments for the photocatalytic degradation of RhB. The catalyst was separated from the reaction suspension by centrifugation, washed with ethanol and water (four times each) and dried in the oven at 100 °C. As shown in Figure 12, NiAl$_2$O$_4$/GQDs retained its photocatalytic activity after four successive experimental runs. A slight decrease was observed after the first use. However, in consecutive runs the photocatalytic activity remained unchanged and retained 96% of its original efficiency.

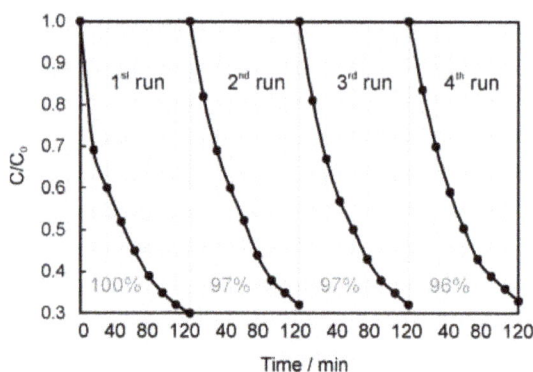

Figure 12. Reusability of the NiAl$_2$O$_4$/GQDs composite.

4. Conclusions

Highly efficient nickel aluminate nanoparticles with spinel crystal structures were successfully synthesized via a simple and cost-effective co-precipitation method. A comprehensive study of the photocatalytic performance of the degradation of different water pollutants, including a series of dyes (i.e., rhodamine B, quinoline yellow, eriochrome black T, methylene blue), phenol and fungicide (thiram) under simulated solar light irradiation was carried out in this study. Moreover, we succeeded in improving the photocatalytic performance of NiAl$_2$O$_4$ by decorating it with GQDs. We presented the physicochemical characterization of the obtained photocatalyst alongside studies of its photocatalytic activity towards rhodamine B and phenol degradations. The mechanism of the photocatalysis in the presence of the NiAl$_2$O$_4$/GQDs composite was studied using the TPA method and a series of scavengers. Hydroxyl radicals were found to play a leading role in the photocatalytic activity of the investigated composite. This work not only offers new insight into the application of the conjunction of the inorganic spinel and the carbon nanostructure (i.e., GQDs), but also provides a simple and highly efficient route for potential water remediation from common pollutants, including dyes and colorless harmful substances. Moreover, the synthesized composite exhibited multifunctionality, which will be further investigated in an upcoming paper.

Author Contributions: Conceptualization, E.R.; methodology, E.R.; TEM investigation, A.B.; investigation other than TEM analysis, E.R. and J.B.; writing—original draft preparation, E.R.; writing—review and editing, E.R., J.B. and A.B.

Funding: We gratefully acknowledge the financial support from the Polish Ministry of Science and Higher Education under subsidy granted to the Faculty of Biology and Chemistry, University of Bialystok for R&D and related tasks aimed at development of young scientists and PhD students and for maintaining the research potential of the Faculty of Biology and Chemistry, University of Bialystok. Diffractometer, IR spectrometer, SEM and TEM microscopes, DSC and TGA instruments, UV–Vis/NIR spectrophotometer and spectrofluorometer were funded by EU, as part of the Operational Programme Development of Eastern Poland, projects Nr POPW.01.03.00-20-034/09-00 and POPW.01.03.00-20-004/11-00.

Acknowledgments: The authors thank A. Wilczewska (University of Bialystok, Poland) for DSC-TGA measurements.

Conflicts of Interest: The authors declare no conflict of interest.

References

1. Ma, G.; Yang, M.; Li, C.; Tan, H.; Deng, L.; Xie, S.; Xu, F.; Wang, L.; Song, Y. Preparation of spinel nickel-cobalt oxide nanowrinkles/reduced graphene oxide hybrid for nonenzymatic glucose detection at physiological level. *Electrochim. Acta* **2016**, *220*, 545–553. [CrossRef]
2. Gaudon, M.; Robertson, L.C.; Lataste, E.; Duttine, M.; Ménétrier, M.; Demourgues, A. Cobalt and nickel aluminate spinels: Blue and cyan pigments. *Ceram. Int.* **2014**, *40*, 5201–5207. [CrossRef]

3. Vitorino, N.M.D.; Kovalevsky, A.V.; Ferro, M.C.; Abrantes, J.C.C.; Frade, J.R. Design of $NiAl_2O_4$ cellular monoliths for catalytic applications. *Mater. Des.* **2017**, *117*, 332–337. [CrossRef]

4. Li, J.; Ren, Y.; Yue, B.; He, H. Ni/Al_2O_3 catalysts derived from spinel $NiAl_2O_4$ for low-temperature hydrogenation of maleic anhydride to succinic anhydride. *Chin. J. Catal.* **2017**, *38*, 1166–1173. [CrossRef]

5. López-Fonseca, R.; Jiménez-González, C.; de Rivas, B.; Gutiérrez-Ortiz, J.I. Partial oxidation of methane to syngas on bulk $NiAl_2O_4$ catalyst. Comparison with alumina supported nickel, platinum and rhodium catalysts. *Appl. Catal. Gen.* **2012**, *437–438*, 53–62.

6. Cesteros, Y.; Salagre, P.; Medina, F.; Sueiras, J. Synthesis and characterization of several $Ni/NiAl_2O_4$ catalysts active for the 1,2,4-trichlorobenzene hydrodechlorination. *Appl. Catal. B Environ.* **2000**, *25*, 213–227. [CrossRef]

7. Farahani, M.D.; Dasireddy, V.D.B.C.; Friedrich, H.B. Oxidative Dehydrogenation of *n*-Octane over Niobium-Doped $NiAl_2O_4$: An Example of Beneficial Coking in Catalysis over Spinel. *ChemCatChem* **2018**, *10*, 2059–2069. [CrossRef]

8. Maddahfar, M.; Ramezani, M.; Sadeghi, M.; Sobhani-Nasab, A. $NiAl_2O_4$ nanoparticles: Synthesis and characterization through modify sol–gel method and its photocatalyst application. *J. Mater. Sci. Mater. Electron.* **2015**, *26*, 7745–7750. [CrossRef]

9. Rahimi-Nasrabadi, M.; Ahmadi, F.; Eghbali-Arani, M. Different morphologies fabrication of $NiAl_2O_4$ nanostructures with the aid of new template and its photocatalyst application. *J. Mater. Sci. Mater. Electron.* **2017**, *28*, 2415–2420. [CrossRef]

10. Tangcharoen, T.; T-Thienprasert, J.; Kongmark, C. Optical properties and versatile photocatalytic degradation ability of MAl_2O_4 (M = Ni, Cu, Zn) aluminate spinel nanoparticles. *J. Mater. Sci. Mater. Electron.* **2018**, *29*, 8995–9006. [CrossRef]

11. Elakkiya, V.; Agarwal, Y.; Sumathi, S. Photocatalytic activity of divalent ion (copper, zinc and magnesium) doped $NiAl_2O_4$. *Solid State Sci.* **2018**, *82*, 92–98. [CrossRef]

12. Akika, F.Z.; Benamira, M.; Lahmar, H.; Tibera, A.; Chabi, R.; Avramova, I.; Suzer, Ş.; Trari, M. Structural and optical properties of Cu-substitution of $NiAl_2O_4$ and their photocatalytic activity towards Congo red under solar light irradiation. *J. Photochem. Photobiol. Chem.* **2018**, *364*, 542–550. [CrossRef]

13. Jayasree, S.; Manikandan, A.; Antony, S.A.; Uduman Mohideen, A.M.; Barathiraja, C. Magneto-Optical and Catalytic Properties of Recyclable Spinel $NiAl_2O_4$ Nanostructures Using Facile Combustion Methods. *J. Supercond. Nov. Magn.* **2016**, *29*, 253–263. [CrossRef]

14. Deraz, N.M. Synthesis and Characterization of Nano-Sized Nickel Aluminate Spinel Crystals. *Int. J. Electrochem. Sci.* **2013**, 5203–5212.

15. Gholami, T.; Salavati-Niasari, M.; Varshoy, S. Electrochemical hydrogen storage capacity and optical properties of $NiAl_2O_4/NiO$ nanocomposite synthesized by green method. *Int. J. Hydrog. Energy* **2017**, *42*, 5235–5245. [CrossRef]

16. Gholami, T.; Salavati-Niasari, M.; Salehabadi, A.; Amiri, M.; Shabani-Nooshabadi, M.; Rezaie, M. Electrochemical hydrogen storage properties of $NiAl_2O_4/NiO$ nanostructures using TiO_2, SiO_2 and graphene by auto-combustion method using green tea extract. *Renew. Energy* **2018**, *115*, 199–207. [CrossRef]

17. Zhang, H.; Hong, H.; Jiang, Q.; Deng, Y.; Jin, H.; Kang, Q. Development of a chemical-looping combustion reactor having porous honeycomb chamber and experimental validation by using $NiO/NiAl_2O_4$. *Appl. Energy* **2018**, *211*, 259–268. [CrossRef]

18. Wei, Y.; Zhang, X.; Wu, X.; Tang, D.; Cai, K.; Zhang, Q. Carbon quantum dots/Ni–Al layered double hydroxide composite for high-performance supercapacitors. *RSC Adv.* **2016**, *6*, 39317–39322. [CrossRef]

19. Iguchi, S.; Teramura, K.; Hosokawa, S.; Tanaka, T. Photocatalytic conversion of CO_2 in water using fluorinated layered double hydroxides as photocatalysts. *Appl. Catal. Gen.* **2016**, *521*, 160–167. [CrossRef]

20. Iguchi, S.; Hasegawa, Y.; Teramura, K.; Hosokawa, S.; Tanaka, T. Preparation of transition metal-containing layered double hydroxides and application to the photocatalytic conversion of CO_2 in water. *J. CO_2 Util.* **2016**, *15*, 6–14. [CrossRef]

21. Khodam, F.; Rezvani, Z.; Amani-Ghadim, A.R. Fabrication of a novel $ZnO/MMO/CNT$ nanohybrid derived from multi-cationic layered double hydroxide for photocatalytic degradation of azo dye under visible light. *RSC Adv.* **2015**, *5*, 19675–19685. [CrossRef]

22. Salehi, G.; Abazari, R.; Mahjoub, A.R. Visible-Light-Induced Graphitic–C$_3$N$_4$@Nickel–Aluminum Layered Double Hydroxide Nanocomposites with Enhanced Photocatalytic Activity for Removal of Dyes in Water. *Inorg. Chem.* **2018**, *57*, 8681–8691. [CrossRef]

23. Yang, M.-Q.; Zhang, N.; Xu, Y.-J. Synthesis of Fullerene-, Carbon Nanotube-, and Graphene-TiO$_2$ Nanocomposite Photocatalysts for Selective Oxidation: A Comparative Study. *ACS Appl. Mater. Interfaces* **2013**, *5*, 1156–1164. [CrossRef] [PubMed]

24. Regulska, E.; Rivera-Nazario, D.M.; Karpinska, J.; Plonska-Brzezinska, M.E.; Echegoyen, L. Enhanced Photocatalytic Performance of Porphyrin/Phthalocyanine and *Bis*(4-pyridyl)pyrrolidinofullerene modified Titania. *ChemistrySelect* **2017**, *2*, 2462–2470. [CrossRef]

25. Regulska, E.; Karpińska, J. Investigation of novel material for effective photodegradation of bezafibrate in aqueous samples. *Environ. Sci. Pollut. Res.* **2014**, *21*, 5242–5248. [CrossRef] [PubMed]

26. Regulska, E.; Karpinska, J. Investigation of Photocatalytic Activity of C$_{60}$/TiO$_2$ Nanocomposites Produced by Evaporation Drying Method. *Pol. J. Environ. Stud.* **2014**, *23*, 2175–2182.

27. Hamadanian, M.; Shamshiri, M.; Jabbari, V. Novel high potential visible-light-active photocatalyst of CNT/Mo, S-codoped TiO$_2$ hetero-nanostructure. *Appl. Surf. Sci.* **2014**, *317*, 302–311. [CrossRef]

28. Zhang, L.-W.; Fu, H.-B.; Zhu, Y.-F. Efficient TiO$_2$ Photocatalysts from Surface Hybridization of TiO$_2$ Particles with Graphite-like Carbon. *Adv. Funct. Mater.* **2008**, *18*, 2180–2189. [CrossRef]

29. Yuan, L.; Yu, Q.; Zhang, Y.; Xu, Y.-J. Graphene–TiO$_2$ nanocomposite photocatalysts for selective organic synthesis in water under simulated solar light irradiation. *RSC Adv.* **2014**, *4*, 15264–15270. [CrossRef]

30. Yadav, H.M.; Kim, J.-S. Solvothermal synthesis of anatase TiO$_2$-graphene oxide nanocomposites and their photocatalytic performance. *J. Alloy Compd.* **2016**, *688*, 123–129. [CrossRef]

31. Chinnusamy, S.; Kaur, R.; Bokare, A.; Erogbogbo, F. Incorporation of graphene quantum dots to enhance photocatalytic properties of anatase TiO$_2$. *Mrs Commun.* **2018**, *8*, 137–144. [CrossRef]

32. Dong, Y.; Shao, J.; Chen, C.; Li, H.; Wang, R.; Chi, Y.; Lin, X.; Chen, G. Blue luminescent graphene quantum dots and graphene oxide prepared by tuning the carbonization degree of citric acid. *Carbon* **2012**, *50*, 4738–4743. [CrossRef]

33. Gupta, B.K.; Kedawat, G.; Agrawal, Y.; Kumar, P.; Dwivedi, J.; Dhawan, S.K. A Novel Strategy to Enhance Ultraviolet Light Driven Photocatalysis from Graphene Quantum Dots Infilled TiO$_2$ Nanotube Arrays. *RSC Adv.* **2015**, *5*, 10623–10631. [CrossRef]

34. Zeng, Z.; Chen, S.; Tan, T.T.Y.; Xiao, F.-X. Graphene Quantum Dots (GQDs) and Its Derivatives for Multifarious Photocatalysis and Photoelectrocatalysis. *Catal. Today* **2018**, *315*, 171–183. [CrossRef]

35. Page, S.E.; Arnold, W.A.; McNeill, K. Terephthalate as a Probe for Photochemically Generated Hydroxyl Radical. *J. Environ. Monit.* **2010**, *12*, 1658–1665. [CrossRef]

36. Liao, Y.; Zhu, S.; Chen, Z.; Lou, X.; Zhang, D. A Facile Method of Activating Graphitic Carbon Nitride for Enhanced Photocatalytic Activity. *Phys. Chem. Chem. Phys.* **2015**, *17*, 27826–27832. [CrossRef]

37. Maniammal, K.; Madhu, G.; Biju, V. X-ray Diffraction Line Profile Analysis of Nanostructured Nickel Oxide: Shape Factor and Convolution of Crystallite Size and Microstrain Contributions. *Phys. E Low-Dimens. Syst. Nanostruct.* **2017**, *85*, 214–222. [CrossRef]

38. Motahari, F.; Mozdianfard, M.R.; Soofivand, F.; Salavati-Niasari, M. NiO nanostructures: Synthesis, characterization and photocatalyst application in dye wastewater treatment. *RSC Adv.* **2014**, *4*, 27654–27660. [CrossRef]

39. Wang, Y.; Kong, W.; Wang, L.; Zhang, J.Z.; Li, Y.; Liu, X.; Li, Y. Optimizing Oxygen Functional Groups in Graphene Quantum Dots for Improved Antioxidant Mechanism. *Phys. Chem. Chem. Phys.* **2019**, *21*, 1336–1343. [CrossRef] [PubMed]

40. Lee, K.M.; Lee, W.Y. Partial Oxidation of Methane to Syngas over Calcined Ni–Mg/Al Layered Double Hydroxides. *Catal. Lett.* **2002**, *83*, 65–70. [CrossRef]

41. Serway, R.A.; Beichner, R.J.; Jewett, J.W. *Physics for Scientists and Engineers*, 5th ed.; Saunders Golden Sunburst Series; Saunders College Publishing: Fort Worth, TX, USA, 2000; ISBN 978-0-03-022654-0.

water

MDPI

Article

Liquid–Liquid Continuous Extraction and Fractional Distillation for the Removal of Organic Compounds from the Wastewater of the Oil Industry

Sonia Milena Vegas Mendoza *, Eliseo Avella Moreno, Carlos Alberto Guerrero Fajardo and Ricardo Fierro Medina

Chemistry Department, National University of Colombia, Cra. 45 No. 26-85, Building 451, 111321 Bogotá, Colombia
* Correspondence: smvegasm@unal.edu.co; Tel.: +57-3504745892

Received: 18 June 2019; Accepted: 11 July 2019; Published: 13 July 2019

check for updates

Abstract: This is the first study to carry out a laboratory-scale assay to assess the potentiality of continuous liquid–liquid extraction with dichloromethane ($CLLE_{DCM}$) and high-power fractional distillation (HPFD) as a treatment to decontaminate the wastewater generated by the petroleum industry (WW). The analytical parameters of treated wastewater (TWW) evidenced a remarkable quality improvement compared to the original WW. $CLLE_{DCM}$–HPFD yielded 92.4%–98.5% of the WW mass as more environmentally friendly water. Compared to the original values determined in the WW, total petroleum hydrocarbon (TPH) decreased by 95.0%–100.0%, and the chemical oxygen demand (COD) decreased by 90.5%–99.9%. Taking into account the yield of the treated water, the amount of pollutant removed, and the risks of each process, the order of the potentiality of these treatments, from highest to lowest, was HPFD > $CLLE_{DCM}$–HPFD > $CLLE_{DCM}$. $CLLE_{DCM}$ treatment alone produced TWW with poorer quality, and the $CLLE_{DCM}$–HPFD sequence involved the greatest consumption of time and energy (0.390–0.905 kWh/kg). $CLLE_{DCM}$-only was the least effective treatment because the TWW obtained failed to comply with the regulations of oil-producing countries.

Keywords: continuous liquid–liquid extraction; fractional distillation; removal of organic compounds; total petroleum hydrocarbon; chemical oxygen demand

---

1. Introduction

The production of crude oil (particularly heavy crude oil) involves the coproduction of process waters. In nondomestic wastewater, the petroleum industry generates a significant load of leachate (WW) with diverse compositions, depending on the geographical location of the exploitation, the crude type, and the method for the extraction of the petroleum, among other factors. The petroleum industry generates large amounts of contaminated WW (Table 1). The United States has official data for the years 2007 and 2012, and Colombia has official data for the year 2015 [1–3].

The WW generated during the production of crude oil usually contains dissolved gases (CO_2, H_2S), salts, suspended solids, radioisotopes, hydrocarbons, and metal ions. In accordance with the guidelines in the Standard Methods for the Examination of Water and Wastewater™ [4] REF, the parameters measured to analyze WW and treated wastewater (TWW) are pH, salinity, electrical conductivity (EC), total alkalinity (TA), total suspended solids (TSSes), total dissolved solids (TDSes), chlorides, total hardness (TH), chemical oxygen demand (COD), biological oxygen demand (BOD), copper, nickel, and other compounds [4].

Table 1. A few indicative records of the amounts of nondomestic wastewater (WW) generated by the oil industry.

Country/Company	Stage or Process (Year)	WW Amount	References
EE. UU.	Up, on, off (2009)	21 billion BW/year	[1,2,5]
Colorado	NR (2016)	300 MMBW/year	[6]
Norway (Norwegian Oil and Gas Association)	Off (2012)	130 million m³/year	[7]
Mexico (Pemex)	NR (2010)	12.04×10^6 m³/year	[8]
Brazil (Petrobras)	Off (2005)	73 million m³/year	[9]
Colombia (Ecopetrol)	Off (2015)	12.45 BWPD	[3]
Oman (Petroleum Development Oman)	Up, On (2008)	800 000 m³/day	[10]
Iran (Marun petrochemical complex)	Do, Ref	200 m³/h	[11]
Iraq (North Rumaila field)	NR	290 000–800 000 BBL/day	[12]
Qatar (Qatari North field)	NR (2014)	23 554 BBLS/D	[13]
United Arab Emirates (Al Ruwais refinery)	Do, Ref (2002–2003)	150 m³/h	[14]
World total	Up	210–300 MMBWD	[3,5,15–17]

Note: Up: Upstream, On: Onshore, Off: Offshore, Do: Downstream, Ref: Refining, NR: Not reported, MMBW: Millions of water barrels, BW o BBL o: Barrels of water, BWPD o BBLS/D: Barrels of water per day.

Different regulations in oil-producing countries establish the maximum limits of pollutants in WW that is discharged into surface water bodies (Table 2). The environment ministries and secretariats of Colombia, Mexico, Brazil, Peru, Venezuela, Asturias (Spain), and China and the World Bank Group (WBG) members' oil-producing countries were consulted to compare differences and similarities between their maximum limits of effluents.

WW can be discharged into surface water bodies and public sewage if it meets the requirements established in the regulatory norms that are in effect in a country, e.g., Spain [18], China [19], Colombia [20], Mexico [21], Brazil [22], Peru [23], Venezuela [24], and the World Bank Group (WBG) [25,26]. As a consequence, the petroleum industry has tested different strategies to regulate pollutant discharge into water bodies, marine water, and water used in households, industry, agriculture, water sports, and in power generation, among other applications.

The results of a detailed review indicated that there are a variety of methods to remove pollutants from wastewater, including (a) physical methods, such as adsorption, cyclones, enhanced flotation [16,27], flocculation [28], and activated carbons [29]; (b) chemical methods, such as precipitation, electrochemical techniques, [16,30], oxidation [27,31], photocatalytic techniques [32], and demulsifiers [33]; (c) biological methods, such as bioreactors [16,30,34]; (d) membrane techniques, such as polymeric, ceramic, or inorganic membranes, microfiltration, ultrafiltration, reverse osmosis, membrane distillation, and nanofiltration [16,27,30,35–39]; and (e) combined or hybrid methods, such as coagulation–flocculation and flotation, biological treatment with activated carbon and reverse osmosis (RO), and bioelectrochemical reactor and coagulation membrane processes [30,35,40,41] for reducing the COD, total petroleum hydrocarbon (TPH), and other contaminants present in petroleum WW.

Among physical treatments, the flocculation with zero-valent iron–ethylenediaminetetraacetic acid (EDTA) and air (ZEA) process and granular-activated carbon removed 92% of COD and 97% of TPH and 72.7%–88.2% of TPH, respectively [28,42,43]; and shaking extraction recovered about 60% of TPH from oil [44]. Of the chemical treatments used, electrocoagulation removed 85.81% of COD [45], and electrochemical removed 85%–96% of COD [46,47].

On the other hand, biological treatment studies have shown that rotating biological contactor (RBC) discs removed 78%–97% of COD and 95%–99% of TPH [48–51], and a membrane bioreactor (MBR) removed 96% of COD [11]. Nevertheless, new technologies have been tested, such as membrane in reverse osmosis (RO) and filtration membranes (which removed 82%–99% of COD [52–55]) and a hybrid microfiltration (MF)/ultrafiltration (UF) process (which removed 94.4%–98.8% of COD [56]).

Table 2. National regulations in some oil-producing countries on permissible maximum limits (PMLs) of pollutants in WW for its environmentally safe discharge.

Analytical Parameter—Units	PML	Norm—Country
pH	6.5–9.0	Supreme Decree 004-2017—Peru [23]
		WBG member's [25,26]
		Resolution 631 of 2015—Colombia [20]
	6.0–9.0	Decree 883 of 1995—Venezuela [24]
		Law 5/2002—Spain [18]
		GB 8978—1996—China [19]
	5.0–9.0	Resolution 430 of 2011—Brazil [22]
	5.0–10.0	NOM-001 of 1996—Mexico [21]
EC (μS)[a]	1000	Supreme Decree 004-2017—Peru [23]
	5000	Law 5/2002—Spain [18]
Chlorides (mg/L)	1.000	Decree 883 of 1995—Venezuela [24]
	1.200	Resolution 631 of 2015—Colombia [20]
	1.200	WBG member's [25]
COD (mg /L O_2)[b]	60	GB 8978—1996—China [19]
	125	WBG member's [25,26]
	180	Resolution 631 of 2015—Colombia [20]
	350	Decree 883 of 1995—Venezuela [24]
	1600	Law 5/2002—Spain [18]
TPH (mg/L)[c]	0.5	Supreme Decree 004-2017—Peru [23]
	5	GB 8978—1996—China [19]
	10	WBG member's [25]
	10	Resolution 631 of 2015—Colombia [20]
	15	Law 5/2002—Spain [18]
	20	Decree 883 of 1995—Venezuela [24]
TSSes (mg/L)[d]	30-35	WBG member's [25,26]
	50	Resolution 631 of 2015—Colombia [20]
	60	NOM-001 of 1996—Mexico [21]
	70	GB 8978—1996—China [19]
	80	Decree 883 of 1995—Venezuela [24]
	$\leq$100	Supreme Decree 004-2017—Peru [23]
	1000	Law 5/2002—Spain [18]

[a] Electrical conductivity, [b] chemical oxygen demand, [c] total petrogenic hydrocarbon, [d] total suspended solids, WBG member's: World Bank Group.

The oil industry uses fractional distillation (FD) or extraction with solvents, and the emphasis of the technique is on obtaining the optimal benefit for the crude oil in refining or recovery operations. However, the use of distillation or extraction as a decontamination treatment of wastewater still requires further research and data. In this respect, catalytic vacuum distillation has been shown to reduce COD by 99% [57]. In addition, distillation has been investigated for the treatment of seawater. On the other hand, an extraction technique used to treat pond sludge removed 67.5% [58] and 40%–60% of COD using the solvents methyl ethyl ketone (MEK) and ethyl acetate (EA) [44].

To date, there have been no publications on the systematic use of continuous liquid–liquid extraction (CLLE) or FD, either individually or in sequence, as preliminary treatments for the decontamination of WW (including decontamination in sedimentation ponds during production).

This work presents and discusses the results of laboratory-scale assays using CLLE with dichloromethane (CLLE$_{DCM}$) and high-power fractional distillation (HPFD), individually and in sequence, on authentic WW to produce TWW of better quality than the original water to illustrate the potentiality of these techniques for decontaminating WW. CLLE$_{DCM}$ and HPFD were specifically chosen due to preliminary trials showing that extraction with DCM removed more contaminants from WW and that distillation while heating low- or medium-potency WW prolonged the time of experimentation without appreciable improvement in the yield or quality of the TWW obtained.

2. Materials and Methods

2.1. Samples and Reagents

According to Protocol No. 1060 in the Standard Methods for the Examination of Water and Wastewater™ [4], 60 L of WW was collected from an oil company located in the Department of Meta (Colombia) and stored at 4 °C to perform physicochemical analyses on the initial WW and to treat it through $CLLE_{DCM}$, HTFD, or $CLLE_{DCM}$–HTFD to obtain the corresponding TWW. Analytical-grade dichloromethane (DCM) from Dongyue Chemical was used in the CLLE.

2.2. Sample Characterization

The initial WW and TWW obtained were analyzed according to the Standard Methods for the Examination of Water and Wastewater ™ [4] to determinate the pH (4500B), salinity (2520A), electrical conductivity EC (2510B), total alkalinity (TA) (2320B), total suspended solids (TSSes) (2540D), total dissolved solids (TDSes) (2540C), chlorides (4110B; D), total hardness (TH) (2340C), and COD (5220B, C, and D). WW and TWW were also analyzed according to the Environmental Protection Agency (EPA, Washington, DC, USA) of the United States to determinate total petroleum hydrocarbon (TPH) (EPA 8015D-EPA 3510C). The data obtained on COD, TPH, TSSes, TDSes, EC, TH, pH, TA, salinity, and chlorides were analyzed to compare water quality on the basis of the regulatory norms in force in the countries discussed in this study.

WW and TWW were also subjected to gas chromatography (GC) on a Hewlett Packard 5890 series II chromatograph operated with ultrahigh-purity nitrogen as the carrier gas, an injector (at 558.15 K), a flame ionization detector (at 593.15 K), an Hewlett-Packard (HP) 3396 series II integrator, and a reverse-phase capillary column DB5 (dimethylpolysiloxane; 30 m × 0.25 mm ID, 0.25 μm) at a temperature ramp of 276.15 K/min from 558.15 to 593.15 K. Phytane and pristine were used as standards.

The estimated energy consumed by the $CLLE_{DCM}$ and HPFD methods was calculated in kilowatt-hours (kWh) from the specific heat of liquid water. The specific heat was determined by measuring the energy required to heat a mass of liquid water (from the initial temperature T_1 to the final temperature T_2) contained in the extract collector of the equipment for $CLLE_{DCM}$ or in the distillation flask of the HPFD assembly at identical conditions to those in the assays for WW. In particular, the heating power and losses of energy through thermal insulation were kept constant.

2.3. Experimental Design and Procedure

2.3.1. $CLLE_{DCM}$

The assembly used for extraction is presented in Figure 1A. Equal and previously weighed volumes of 0.35 L of DCM and WW were subjected to processing in the extraction chamber. From the extract collector agitated at 800 min^{-1} at a temperature of <312.75 K (heated by a bath at 333.15 K), cold extract batches were collected during the process at 1200, 2400, 3600, 5400, 7200, 10,800, and 14,400 s after the beginning of solvent condensation on the extraction chamber after the extract collector was replenished with a volume of DCM equal to that of the collected extract. The initial WW, each batch of extract, solvent-free extract obtained by distillation in a rotary evaporator and by vacuum-drying (333.15 K, 2933.1 Pa), and the final raffinate were weighed to within a 0.0001-g accuracy. In this way, $CLLE_{DCM}$ and $CLLE_{DCM}$–HPFD were repeated four and six times, respectively. Estimated energy consumption and operation time were measured from the onset of heating the system until the last condensation drop fell into the extraction chamber for the extract batch at 1200 s and from the first to the last drop that fell for all other batches of extract. The quality improvement in the TWW obtained was assessed through the analytical parameters of the sample.

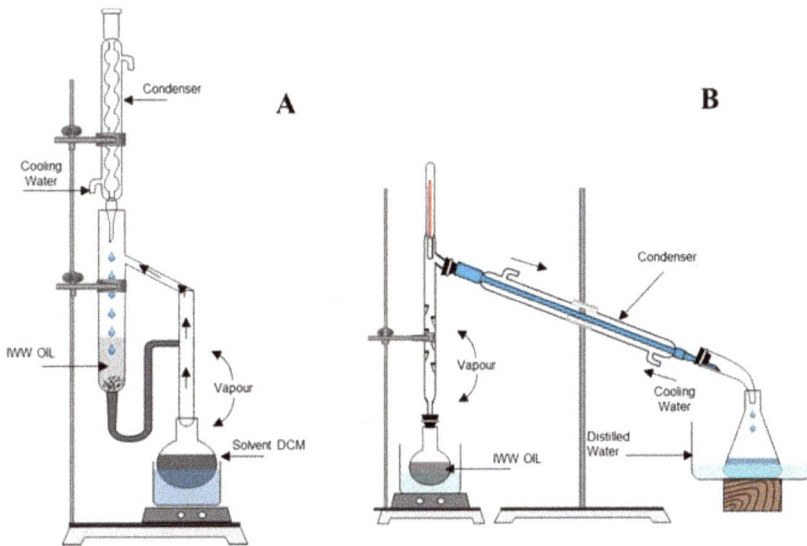

Figure 1. (**A**) Laboratory assembly used for the continuous liquid–liquid extraction with dichloromethane (CLLE$_{DCM}$) and (**B**) high-power fractional distillation (HPFD) of the wastewater (WW).

2.3.2. HPFD

Approximately 0.30 L of previously weighed WW or raffinate from CLLE$_{DCM}$ (TWW) was deposited into the distillation flask of the assembly (shown in Figure 1B) and distilled using a Vigreux fractionation column and a heating mantle at its highest potency. Distillate fractions were collected at 357.15–363.15 K (1–5 mL, head), 363.15–364.15 K (0.24–0.25 L, body), and 364.15–365.15 K (1–5 mL, tail). The feed, WW, or raffinate distillate fractions and each final distillation bottom were weighed to within a 0.0001-g accuracy. In this way, HPFD and CLLE$_{DCM}$–HPFD were repeated 16 and 6 times, respectively. Estimated energy consumption and operation time were measured from the beginning of system heating to the collection of the last drop of distilled TWW. The quality improvement in the TWW obtained was assessed through the analytical parameters of the sample.

2.4. Statistical Analysis

Masses, times, yields, and estimated energies that were recorded during the repetitions of CLLE$_{DCM}$, HPFD, and CLLE$_{DCM}$–HPFD of WW were subjected to descriptive and inferential statistical analysis using parametric and nonparametric statistics according to the results of the tests for data normality (Shapiro–Wilk test and χ^2), variance homogeneity (Bartlett, Levene, or Welch test for data not adjusted to normality), and analysis of variance (ANOVA and Tukey's test if ANOVA was significant). All analyses were performed using Statistical Package for the Social Sciences software (IBM SPSS® version 25.0) [59], at a level of significance of 5%.

2.5. Calculation of Estimated Energy Consumption

The estimated energy consumption (ΔE_s) of each system for the CLLE$_{DCM}$ or HPFD of a determined mass of WW during the operation time (t_s) was calculated from the energy (ΔE) required to heat a mass (m) of liquid water with a determined specific heat (s_h) from an initial temperature (T_1) to a final temperature (T_2) for a measured time (t) using the same system. Operations were assumed to have identical conditions as much as possible. Thus, the following equation was used:

$$\Delta E_s = m(s_h)(T_2 - T_1)\left(\frac{t_s}{t}\right) = \Delta E\left(\frac{t_s}{t}\right), \tag{1}$$

where ΔE_s or ΔE can be expressed as kWh equivalent to 3.6×10^6 J.

3. Results and Discussion

3.1. Guidelines on the Permissible Limits of Pollutant Discharge from WW

The maximum permissible limits of COD and TPH in the norms of China, Peru, WBG members, and Colombia are stricter than those in other countries. In Colombia, the norms regulating the quality of water discharged into water bodies include more analytical parameters than those of Venezuela and Asturias (Spain), and they have less rigorous limits for TSSes, TPH, and COD than do WBG members (Table 2). Regulations in Brazil and Mexico do not require the evaluation of most of the analytical parameters included in the norms of other countries.

3.2. Physicochemical Analysis of WW

Previous studies [28,43,60,61] have presented results on the analytical characterization of WW but have not specified the type of operation or the production stage in which the samples were collected. Table 3 shows data from the physicochemical analysis of the WW sampled from the sedimentation pond of the production area of an oil company in Colombia. The values of the physicochemical parameters in the WW were analyzed and found to exceed the permissible limits for effluent discharge into surface water bodies according to any of the norms of the countries mentioned in this study.

Table 3. Physicochemical properties of the nondomestic wastewater (WW) and the treated wastewater (TWW) obtained through CLLE$_{DCM}$, HPFD, or CLLE$_{DCM}$-HPFD for samples of WW from the sedimentation pond of a petroleum production area.

Test	Permissible Limit by Norm	WW	TWW Obtained by		
			HPFD	CLLE$_{DCM}$	CLLE$_{DCM}$–HPFD
TPH (mg/L) [b]	0.5 [23]	69,287	<0.007	40.88	2.05
COD (mg /L O$_2$) [c]	60.0 [19]	83,100	324	7900	108
pH	6.5–9.00 [23]	6.08 (19.7) [a]	5.22 (23.8) [a]	9.36 (17.6) [a]	7.50 (19.5) [a]
Salinity (mg/L)	NR [18–26]	7410 (19.7) [a]	59.9 (23.9) [a]	346 (17.6) [a]	100 (19.5) [a]
EC (µS/cm)	1000 [23]	12,810 (19.7) [a]	132.3 (23.5) [a]	726 (17.6) [a]	226 (19.5) [a]
TA (mg /L CaCO$_3$) [d]	AR [20]	1880	58	-	88
Chlorides (mg/L)	1000.0 [24]	3203.5	<2.0	-	2.0
TH (mg /L CaCO$_3$) [e]	NR [20]	1380	<4	-	<4
TDSes (mg/L) [f]	NR [18–26]	7788	-	-	-
TSSes (mg/L) [g]	30 [26]	333	-	-	-

NR: Not required, AR: analysis and report required. [a] Temperature (°C), [b] total petrogenic hydrocarbon, [c] chemical oxygen demand, [d] total alkalinity, [e] total hardness, [f] total dissolved solids, [g] total suspended solids.

The peaks corresponding to aliphatic isoprenoid hydrocarbons (pristane (n-C17) and phytane (n-C18)) in the GC chromatogram of the WW (Figure 2A) are indicative of TPH of marine origin [62] (contaminants in the sample). In the GC chromatogram of WW, the unimodal distribution of peak intensities corresponded mainly to n-alkanes of low molecular weight (Cn, n ≤ 25), which ranged from n-C10 to n-C32, from n-C13 to n-C31, and from n-C17 to n-C31 [63]. These results were evidence of the presence of aliphatic and aromatic contaminants.

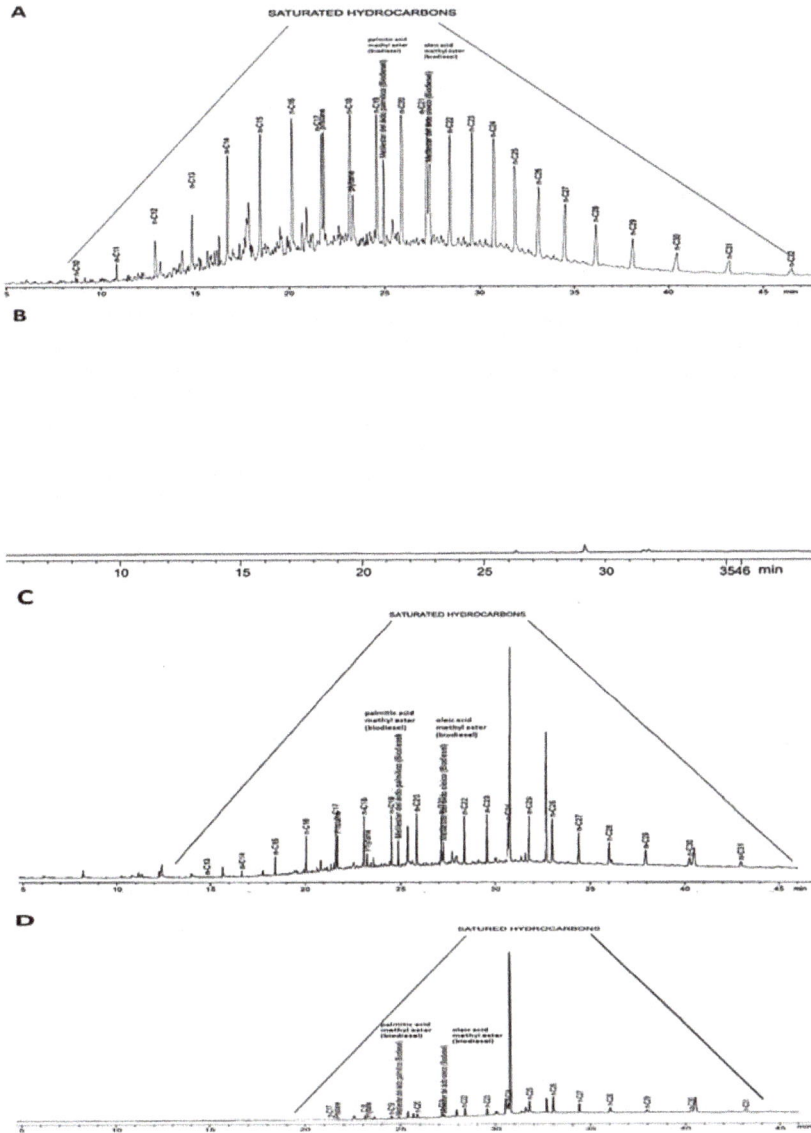

Figure 2. Gas chromatography (GC) (**A**) nondomestic wastewater (WW), (**B**) the treated wastewater (TWW) obtained through HPFD, (**C**) the TWW obtained through CLLE$_{DCM}$, and (**D**) a raffinate (partially treated water) produced through CLLE$_{DCM}$–HPFD of the nondomestic wastewater (WW). Markers: n-alkanes, pristine, and phytane.

In the WW and GC chromatograms of TWW (Figure 2A–D), pentacyclic triterpenes (hopanes) (n-C27–C32) were identified. They included steranes (n-C27–C29), with a predominance of n-C27 [64], and an n-C30/n-C29 ratio higher than 1.0 [65]. All of these species were remnants of pollutants present in the original WW, and they were found in notably minor concentrations in the TWWs obtained by using any of the tested treatments (CLLE$_{DCM}$, HPFD, or CLLE$_{DCM}$–HPFD).

3.3. Physicochemical Analysis of TWW by CLLE$_{DCM}$, HTFD, or CLLE$_{DCM}$–HTFD

Columns 4–6 in Table 3 show the values of the physicochemical parameters characterizing each TWW obtained by performing CLLE$_{DCM}$, HPFD, or CLLE$_{DCM}$–HPFD on WW. Except for the pH, the other values of the physicochemical parameters in the analysis of the TWW generally decreased significantly. Judging by the appearance of the TWW obtained and by the decrease in TPH, COD, and chloride, the HPFD and CLLE$_{DCM}$–HPFD treatments were more effective than CLLE$_{DCM}$ in the removal of the pollutant load from WW. Nonetheless, CLLE$_{DCM}$ removed most of the TPH (>99%) and diminished the COD in the WW by more than 90% (Table 4).

Table 4. Decrease of total petroleum hydrocarbon (TPH) and chemical oxygen demand (COD) by HPFD, CLLE$_{DCM}$, or CLLE$_{DCM}$–HPFD (on a laboratory scale) in the nondomestic wastewater sampled in the sedimentation pond of the petroleum production area (WW).

Treatment	TPH (mg/L)		ΔTPH (%)	COD (mg O$_2$/L)		ΔCOD (%)
	WW	TWW		WW	TWW	
HPFD	69,287	<0.007 **	99.9	83,100	324 **	99.6
CLLE$_{DCM}$	69,287	40.88 *	99.9	83,100	7900 *	90.5
CLLE$_{DCM}$–HPFD	69,287	40.88 *	99.9	83,100	7900 *	90.5
CLLE$_{DCM}$–HPFD	40.88 *	2.05 **	95.0	7900 *	108 **	98.6
CLLE$_{DCM}$–HPFD	69,287	2.05 **	100.0	83,100	108 **	99.9

TWW = treated water obtained, ΔTPH = decrease of TPH, ΔCOD = decrease of COD, * raffinate, ** distillate.

Similar or higher decreases in COD and TPH from decontamination treatments of wastewater have been reported in the literature by only a few other studies using more expensive and sophisticated treatments that require special equipment and more controlled conditions than those of HTFD or CLLE$_{DCM}$ [42,66–68].

The pH of the TWW obtained by HTFD or CLLE$_{DCM}$ did not comply with the requirement of most regulations (Tables 2 and 3). These pH values are explainable as an effect of the treatments changing the concentrations of chemical species in acid–base equilibria that were present as pollutants in the WW [69,70]. An example of such a mechanism is considered in Scheme 1.

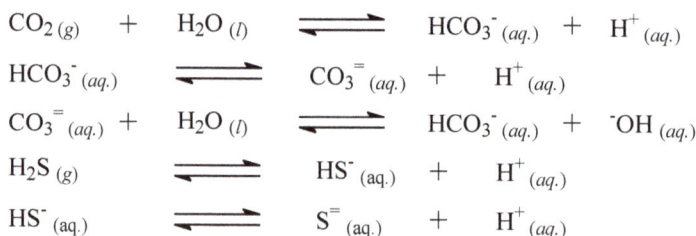

$$CO_{2\,(g)} \;+\; H_2O_{\,(l)} \rightleftharpoons HCO_3^-{}_{(aq.)} \;+\; H^+{}_{(aq.)}$$

$$HCO_3^-{}_{(aq.)} \rightleftharpoons CO_3^{=}{}_{(aq.)} \;+\; H^+{}_{(aq.)}$$

$$CO_3^{=}{}_{(aq.)} \;+\; H_2O_{\,(l)} \rightleftharpoons HCO_3^-{}_{(aq.)} \;+\; {}^-OH_{(aq.)}$$

$$H_2S_{\,(g)} \rightleftharpoons HS^-{}_{(aq.)} \;+\; H^+{}_{(aq.)}$$

$$HS^-{}_{(aq.)} \rightleftharpoons S^{=}{}_{(aq.)} \;+\; H^+{}_{(aq.)}$$

Scheme 1. Acid–base equilibria between carbonate ions (CO$_3^{=}$$_{(aq.)}$), bicarbonate ions (HCO$_3^-$$_{(aq.)}$), and carbon dioxide (CO$_{2(g)}$), and between sulfide ions (S$^{=}$$_{(aq.)}$), disulfide ions (S$^{=}$$_{(aq.)}$), and hydrogen sulfide (H$_2$S$_{(g)}$).

With the HPFD treatment, CO$_{2(g)}$ and H$_2$S$_{(g)}$ went from the hot bottom to the steam phase. Cooling these compounds resulted in their redissolution and concentration, and the pH of the distillate decreased (pH 5.22). The solubility of CO$_{2(g)}$ and H$_2$S$_{(g)}$ in the extract decreased as the extract warmed, and these gases were then lost from the gas phase during solvent evaporation–condensation cycles in the CLLE$_{DCM}$ treatment. The ions in equilibrium, as dissolved species, were concentrated in the raffinates, and the pH increased (pH 9.36).

The raffinates of the CLLE$_{DCM}$ treatment with scarce CO$_{2(g)}$ and H$_2$S$_{(g)}$ products were subjected to HPFD, which produced distillate with an appropriate pH (pH 7.50). In other words, the TWW obtained by CLLE$_{DCM}$–HPFD complied with the current legal requirements of the countries examined in this study [18–26].

The conductivity, salinity, and alkalinity of the raffinates obtained by CLLE$_{DCM}$ or CLLE$_{DCM}$–HTFD directly related to the high concentrations of Cl⁻, Na⁺, and Ca²⁺ in the original WW [69,70]. These characteristics are unfavorable for the quality of the TWW yielded from these treatments.

The peak intensities in the GC chromatograms of the TWW product of CLLE$_{DCM}$, HPFD, or CLLE$_{DCM}$–HPFD treatments of WW (Figure 2B–D) evidenced a lower TPH content than in the starting WW (Figure 2A). The peaks between n-C13 and n-C31 in their GC chromatograms decreased (Figure 2C,D) or practically disappeared (Figure 2B). The persistence of the peaks corresponding to pristane and phytane ($t_R \approx 21.7$ and 23.2 min) and the increase in the pristane/phytane ratio, particularly in the GC chromatograms of the TWW obtained as raffinates (Figure 2C,D), indicated the presence of n-alkanes, as well as methyl palmitate and methyl oleate (residuals of biodiesel, $t_R \approx 24.9$ and 27.1 min), in the TWW.

Since their peaks were absent in the chromatogram of the extracts, the transfer of those species from the feed to the extracts in the CLLE$_{DCM}$ process must have been incomplete. The similarities between the GC chromatograms of WW and the raffinates confirmed the common origin of the pollutants found in those samples (Figure 2A,C,D). The prevalence of a few of such peaks in the GC chromatograms of the distillates obtained by HPFD or CLLE$_{DCM}$–HPFD indicated that a tiny fraction of the contaminant load was transferred from the WW to distillates during the process. The absence of the majority of such peaks in the GC chromatogram of the distillate showed the potentiality of HPFD for decontaminating WW (Figure 2B).

3.4. Operation Time, Energy Consumption, and Performance of CLLE$_{DCM}$, HPFD, and CLLE$_{DCM}$–HPFD

Table 5 shows the average duration and the average mass of the TWW relative to the yields obtained. The average energy consumption (kWh) and time (h) required to obtain a unit of mass (kg) of the respective TWW are also shown, which were calculated after being subjected to *n* measured durations and masses according to the statistical treatment described in Section 2.4. Such measures and estimates (Tables 4 and 5) were established at the given laboratory conditions described in Section 2.3.

Table 5. Yields and average estimated consumption of time and energy per unit mass of treated water (TWW) obtained by CLLE$_{DCM}$, HPFD, or CLLE$_{DCM}$–HPFD (on a laboratory scale) of nondomestic wastewater (WW) of sedimentation ponds in the petroleum production stage.

Treatment	n	Average Duration (min)	TWW Mass (g) $\overline{m} \pm$ SD	Estimated Consumption		Yield (%)
				kWh/kg	h/kg	
HPFD	16	51 ± 3.4	247.7 ± 64.0 [a]	0.360 ± 0.125 [a]	3.4 ± 0.8	97.3
CLLE$_{DCM}$	28	240 ± 0.1	342.5± 15.0 [b]	0.390 ± 0.018 [b]	11.7 ± 0.0	98.5
CLLE$_{DCM}$–HPFD	42	240 ± 0.1	334.5 ± 7.0 [b]	0.400 ± 0.009[b]	12.0 ± 0.2	98.5
CLLE$_{DCM}$–HPFD	6	50 ± 1.6	244.1 ± 18.5 [a]	0.358 ± 0.029 [a]	3.4 ± 0.2	94.1
CLLE$_{DCM}$–HPFD	6	290 ± 1.7	244.1 ± 18.5 [a]	0.905 ± 0.007 [c]	19.8 ± 0.5	92.4

n: Number of data points; $\overline{m}$: Average mass; SD: Standard deviation; g: Grams; [a] distillate average mass; [b] raffinate average mass; [c] average mass of raffinate plus distillate.

The CLLE$_{DCM}$ treatment recovered 98.5% of the initial WW mass, indicating an approximate yield of 50.0 kg/kWh of TWW at a rate of 0.1 kg/h. CLLE$_{DCM}$ produced yellowish and opalescent water with a pH of 9.36, and TPH and COD decreased by 99% and 90.5%, respectively, compared to the starting WW: That is, CLLE$_{DCM}$ did not achieve percentages of removal of TPH, COD, TH, TDSes, and TSSes that would allow the TWW obtained by this method to meet the regulatory requirements of the countries in this study.

The inferential statistical analysis (Shapiro–Wilk test) of the mean weight of solvent-free extracts collected at 20, 40, 60, 90, 120, 180, and 240 min in the four CLLE$_{DCM}$ replicates (28 values) and six CLLE$_{DCM}$–HPFD replicates (42 values) indicated that these data fit a normal distribution ($p > 0.05$)

(Table 6A). The results of the χ^2 test indicated that the mean weights of the solvent-free extracts collected at different time points in these replicates were independent ($p > 0.05$) (Table 7).

Table 6. Statistics of the (**A**) normality test (Shapiro–Wilk) and of (**B**) multiple comparisons (HSD Tukey) applied to the data from the $CLLE_{DCM}$ and $CLLE_{DCM}$–HPFD of industrial wastewater (WW).

A		Statistics of Normality		B		Differences of Averages	
$CLLE_{DCM}$		**$CLLE_{DCM}$–HPFD**		**Time (min) ***	**$CLLE_{DCM}$**	**$CLLE_{DCM}$–HPFD**	
four data points		six data points					
ST [a]	Sig. A. Bil [b]	ST [a]	Sig. A. Bil [b]				
		0.86	0.139	40	0.21	0.37 **	
0.89	0.266	0.86	0.155	60	0.65 **	0.84 **	
0.75	0.130	0.86	0.156	90	1.27 **	1.36 **	
0.88	0.212	0.85	0.112	120	1.51 **	1.80 **	
0.92	0.470	0.87	0.188	180	1.69 **	1.88 **	
		0.90	0.359	240	1.75 **	1.89 **	

[a] Statistical test; [b] Sig. A. Bil: bilateral asymptotic significance; * in relation to a minimum processing time of 20 min; ** the difference of averages in relation to the target is significant at the 0.05 level, HSD: Honestly-significant-difference.

Table 7. Sample of results of the inferential and descriptive statistics of the application of $CLLE_{DCM}$ and $CLLE_{DCM}$–HPFD in decontaminating the industrial wastewater (WW) of a sediment pool.

Treatment/Test		χ^2 by Pearson [c]	Bartlett's Sphericity [d]	Levene Test [d]	Welch Test [d]	ANOVA (F) [e]
$CLLE_{DCM}$	EP [a]	84.0	35.7	3.6	66.0	47.9
	SAB [b]	0.388	0.000	0.014	0.000	0.000
$CLLE_{DCM}$–HPFD	EP [a]	210.0	52.9	9.2	122.6	110.9
	SAB [b]	0.391	0.000	0.000	0.000	0.000

[a] Statistical test; [b] sig. asymptotic (bilateral); [c] from 24 to 34 boxes (100.0%) have expected frequencies less than 5; [d] homogeneity of variance; [e] univariate analysis of variance.

The Welch test ($p < 0.05$) was applied in cases of violation of the homoscedasticity assumption in the Levene and Bartlett tests. There were no significant differences in the mean weights of the extracts, and this is a requirement for the ANOVA test. The ANOVA analysis indicated that there were significant differences ($p < 0.05$) in the mean weight of the extracts collected at different time points (Table 7).

The results of Tukey's test demonstrated that the data on the mean weight of the solvent-free extracts from four $CLLE_{DCM}$ replicates and five $CLLE_{DCM}$–HPFD replicates were significantly different ($p < 0.05$) and that the highest percentage of contaminants was removed in the first 40 min using $CLLE_{DCM}$ and in the first 20 min using $CLLE_{DCM}$–HPFD (Table 6B).

The HPFD treatment recovered 94.1%–97.3% of the initial WW mass, indicating an approximate yield of 2.8 ± 0.7 kg/kWh at a rate of 0.3 ± 0.1 kg/h. HPFD produced colorless, transparent water with a pH of 5.22 (a low value), and TPH and COD decreased by 95.0%–99.9% and 98.6%–99.6%, respectively, compared to the starting WW. The TWW obtained by HPFD presented the lowest residual values of the analytical parameters for TPH, EC, TA, TH, TDSes, TSSes, chlorides, and salinity measured in the original WW analysis (Table 3 and Figure 2B).

The $CLLE_{DCM}$–HPFD treatment recovered 92.4% of the initial WW mass, indicating an approximate yield of 9.1 ± 0.2 kg/kWh of TWW at a rate of 0.05 kg/h. $CLLE_{DCM}$–HPFD produced colorless and transparent water with a pH of 7.50, and TPH and COD decreased by 100.0% and 99.9%, respectively, compared to the starting WW. The TWW obtained by $CLLE_{DCM}$–HPFD met the requirements of current regulations in WBG countries, except for Peru and China, but presented appreciable remnants of salinity, TH, TDSes, and TSSes. Among the three treatments, this method also had the highest estimated average consumption of time and energy required.

Nonetheless, energy consumption decreased and performance slightly increased (98.5%) as the operation time of HPFD increased (Table 5). Therefore, it is still better to produce TWW with the quality achieved by individual HPFD process than wastewater with slightly better quality that meets the regulations of the analyzed countries and WBG member's for effluent discharge (Tables 2 and 3) at a relatively lower yield (92.4%), longer operation time, and higher energy consumption (using $CLLE_{DCM}$–HPFD) (Table 5).

A duration of 100 min for the $CLLE_{DCM}$ operation, as either an individual treatment or as part of the $CLLE_{DCM}$–HTFD sequence, is recommended in order to remove a relatively high fraction of the pollutants from WW with a reasonable minimal consumption of time and energy (Figure 3).

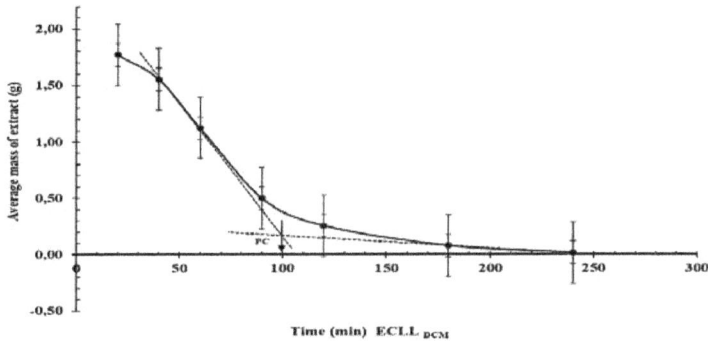

Figure 3. Average mass of solvent-free extracts obtained in relation to the time of process during the $CLLE_{DCM}$ of WW.

The recommended total duration of operation for the other treatments (HPFD or $CLLE_{DCM}$–HPFD) and the yields of TWW depend on the boiling regime, particularly at the end of the process. When the boiling was affected by the concentration or even the crystallization of poorly soluble contaminants in the distillation bottom, material projections could propagate from the bottom to the distillate collector.

Obviously, the yields, and probably the quality, of the TWW obtained by using any of these treatments can be optimized by using better-controlled conditions than those used in this first instance of laboratory-scale experiments. The laboratory conditions in this study that can be optimized include the control of losses due to handling, heating, the use of a vacuum, temperature control, and improvements in the energy transfer and utilization in each treatment system. Such optimization is outside the scope of this work, but it could be of some engineering interest to take advantage of the demonstrated potential of these treatments for decontaminating the WW generated by the petroleum industry or other sources.

4. Conclusions

This work constitutes the first systematic assay on a laboratory scale determining the potential for using $CLLE_{DCM}$, HPFD, or $CLLE_{DCM}$–HPFD as treatments to decontaminate wastewater generated in sedimentation ponds in the production stage of the petroleum industry (WW). The results of using these treatments demonstrated significant yields (94.1%–97.3% by HPFD, 98.5% by $CLLE_{DCM}$, and 92.4% by $CLLE_{DCM}$–HPFD) of water, in which decreases in TPH ranged from 95.0% (HPFD) to 100% ($CLLE_{DCM}$–HPFD), decreases in COD ranged from 90.5% ($CLLE_{DCM}$) to 99.9% ($CLLE_{DCM}$–HPFD), and most of the other pollutants decreased compared to the starting WW: That is, water treated by HPFD and $CLLE_{DCM}$–HPFD met the requirements of the permissible limits of TPH and COD in national and international regulations, including Colombia and China.

Among the tested treatments to decontaminate WW, HPFD was a more effective treatment than $CLLE_{DCM}$–HPFD or $CLLE_{DCM}$ because it produced at least a 92.4% yield of water with an acceptable quality (only slightly less than the quality achieved by $CLLE_{DCM}$–HPFD and higher than that obtained using CLLE$_{DCM}$). Furthermore, HPFD neither demands the long process time or large energy consumption typical of CLLE$_{DCM}$–HPFD nor produces the low water quality obtained by CLLE$_{DCM}$, and its pH of 5.22 can be easily adjusted to meet all the requirements of the norms.

The demonstrated potential of HPFD could be attractive as an engineering optimization study. If the scarcity of water of good quality in the world becomes critical, humanity will need to meet the strictest regulations on the disposal and environmentally safe use of wastewater to satisfy water demand and thus preserve life on the planet, which will supersede concerns for economic and energy costs. The factors that currently render the use of HPFD prohibitive for the decontamination of WW will be comparatively minor problems among the others that engineers will have to solve in the future.

Author Contributions: Conceptualization, S.M.V.M. and E.A.M.; methodology, S.M.V.M. and E.A.M.; software, S.M.V.M.; validation, C.A.G.F. and R.F.M.; formal analysis, S.M.V.M., E.A.M. and C.A.G.F. investigation, S.M.V.M.; resources, S.M.V.M., E.A.M., C.A.G.F. and R.F.M. data curation, S.M.V.M. and E.A.M.; writing—original draft preparation, S.M.V.M. and E.A.M.; writing—review and editing, C.A.G.F. and R.F.M.; visualization, C.A.G.F.; supervision, C.A.G.F.; project administration, S.M.V.M.; funding acquisition, S.M.V.M.

Funding: The authors would like to thank the Administrative Department of Science, Technology, and Innovation (COLCIENCIAS) for the approval of her doctoral scholarship in chemistry at the National University of Colombia-Bogotá (COLCIENCIAS 647).

Acknowledgments: S.M.V.M. is grateful to the Administrative Department of Science, Technology, and Innovation (COLCIENCIAS) for the approval of her doctoral scholarship in chemistry at the National University of Colombia-Bogotá (COLCIENCIAS 647). All authors are grateful to the Chemistry Department of the National University of Colombia at Bogotá for providing the facilities and aids that allowed for the developments of this work as part of the doctoral thesis titled "Application of Continuous Liquid–Liquid Extraction and Fractional Distillation as Primary Treatments to Decontaminate Industrial Wastewaters Generated by the Petroleum Industry".

Conflicts of Interest: The authors declare no conflict of interest.

References

1. Veil, J. Produced Water Volumes and Management Practices in 2012. Available online: http://www.veilenvironmental.com/publications/pw/prod_water_volume_2012.pdf (accessed on 4 December 2017).

2. Dickhout, J.M.; Moreno, J.; Biesheuvel, P.M.; Boels, L.; Lammertink, R.G.; De Vos, W.M. Produced water treatment by membranes: A review from a colloidal perspective. *J. Colloid Interface Sci.* **2017**, *487*, 523–534. [CrossRef] [PubMed]

3. Mesa, S.L.; Orjuela, J.M.; Ramírez, A.T.; Herrera, J.A. Revisión del panorama actual del manejo de agua de producción en la industria petrolera colombiana. *Gestión y Ambiente* **2018**, *21*, 87–98. [CrossRef]

4. Rice, E.; Bridgewater, L. *Standard Methods for the Examination of Water and Wastewater*, 22nd ed.; American Public Health Association (APHA), American Water Works Association (AWWA), Water Environment Federation (WEF), Eds.; American Public Health Association: Washington, DC, USA, 2012; pp. 38–46, ISBN 978-08-7553-013-0.

5. Al-Ghouti, M.A.; Al-Kaabi, M.A.; Ashfaq, M.Y.; Dana, A.D. Produced water characteristics, treatment and reuse: A review. *J. Water Process Eng.* **2019**, *28*, 222–239. [CrossRef]

6. Dolan, F.C.; Cath, T.Y.; Hogue, T.S. Assessing the feasibility of using produced water for irrigation in Colorado. *Sci. Total Environ.* **2018**, *640–641*, 619–628. [CrossRef] [PubMed]

7. Bakke, T.; Klungsøyr, J.; Sanni, S. Environmental impacts of produced water and drilling waste discharges from the Norwegian offshore petroleum industry. *Mar. Environ. Res.* **2013**, *92*, 154–169. [CrossRef]

8. Martel-Valles, J.F.; Foroughbakchk-Pournavab, R.; Benavides-Mendoza, A. Produced waters of the oil industry as an alternative water source for food production. *Rev. Int. Contam. Ambient.* **2016**, *32*, 463–475. [CrossRef]

9. Gabardo, I.T.; Platte, E.B.; Araujo, A.S.; Pulgatti, F.H. Evaluation of produced water from Brazilian offshore platforms. In *Produced Water*; Springer: Berlin/Heidelberg, Germany, 2011; pp. 89–113, ISBN 978-1-4614-0046-2.

10. Breuer, R.; Al-Asmi, S.R. Nimr Water Treatment Project—Up Scaling A Reed Bed Trail To Industrial. In Proceedings of the SPE International Conference on Health, Safety and Environment in Oil and Gas Exploration and Production, Rio de Janeiro, Brazil, 10 November 2018; p. 11.

11. Bayat, M.; Mehrnia, M.R.; Hosseinzadeh, M.; Sheikh-Sofla, R. Petrochemical wastewater treatment and reuse by MBR: A pilot study for ethylene oxide/ethylene glycol and olefin units. *J. Ind. Eng. Chem.* **2015**, *25*, 265–271. [CrossRef]

12. Kuraimid, Z.K. Treatment of Produced Water in North Rumela Oil Field for Re-Injection Application. In Proceedings of the SPE Kuwait Oil and Gas Show and Conference, Kuwait City, Kuwait, 10 November 2018; p. 12.

13. AlKaabi, M.A. Enhancing Produced Water Quality using Modified Activated Carbon. Master's Thesis, Qatar University, Doha, Qatar, 2016.

14. Al Zarooni, M.; Elshorbagy, W. Characterization and assessment of Al Ruwais refinery wastewater. *J. Hazard. Mater.* **2006**, *136*, 398–405. [CrossRef] [PubMed]

15. Alsari, A.; Ibrahim, H.; Okasha, A.; Aboabboud, M.M. Treatment System of Produced Water with Supercritical Carbon Dioxide. *IPCBEE* **2014**, *68*, 40–44. [CrossRef]

16. Jiménez, S.; Micó, M.M.; Arnaldos, M.; Medina, F.; Contreras, S. State of the art of produced water treatment. *Chemosphere* **2018**, *192*, 186–208. [CrossRef] [PubMed]

17. Liang, Y.; Ning, Y.; Liao, L.; Yuan, B. Chapter Fourteen—Special Focus on Produced Water in Oil and Gas Fields: Origin, Management, and Reinjection Practice. In *Formation Damage During Improved Oil Recovery*; Yuan, B., Wood, D.A., Eds.; Gulf Professional Publishing: Houston, TX, USA, 2018; pp. 515–586, ISBN 978-0-12-813782-6.

18. Official Journal of the Principality of Asturias (Boletín Oficial del Principado de Asturias). No. 5/2002, of 3 June 2002, on Discharges of Industrial Wastewater to Public Sanitation Systems. *Off. J. Princ. Astur.* **2002**, 26180–26187. Available online: https://www.boe.es/eli/es-as/l/2002/06/03/5 (accessed on 26 December 2018).

19. State Environmental Protection Administration of the People's Republic of China. GB 8978-1996, Integrated Wastewater Discharge Standard. 1996. Available online: http://english.mee.gov.cn/Resources/standards/water_environment/Discharge_standard/200710/t20071024_111803.shtml (accessed on 26 April 2018).

20. Official Journal of the Republic of Colombia (Diario Oficial de la Republica de Colombia). Resolution No. 631 of 2015; Establishes the Maximum Permitted Limits Values are Established in the Point Discharges to Surface Water Bodies and Sewage Systems Public and Other Provisions are Issued. Official Journal 49.486; 2015. Available online: http://svrpubindc.imprenta.gov.co/diario/view/diarioficial/consultarDiarios.xhtml (accessed on 26 March 2018).

21. Official Journal of the United Mexican States (Diario Oficial de la Federación de los Estados Unidos Mexicanos). NOM-001-SEMARNAT-1996. Establishes the Maximum Permitted Limits of Pollutants in the Wastewater Discharges into Water and onto National Property. 2003. Available online: http://www.ordenjuridico.gob.mx/Documentos/Federal/wo69205.pdf (accessed on 26 March 2018).

22. Official Journal of the Federative Republic of Brazil (Diário Oficial da República Federativa do Brasil). Resolution No. 430 of 2011. Establishes the Conditions and Patterns of Effluent Release, Complements and Alters Resolution No. 357, of March 17, 2005, of the National Council of the Environment-CONAMA. Official Journal No. 92. 2011. Available online: https://www.jusbrasil.com.br/diarios/26738562/pg-89-secao-1-diario-oficial-da-uniao-dou-de-16-05-2011 (accessed on 12 April 2018).

23. Official Journal of the Republic of Peru (Diario Oficial de la Republica de Perú). Supreme Decree No. 004-2017, Establishes the Environmental Quality Standards (ECA) for Water and Establishes Complementary Provisions. 2017. Available online: https://busquedas.elperuano.pe/normaslegales/aprueban-estandares-de-calidad-ambiental-eca-para-agua-y-e-decreto-supremo-n-004-2017-minam-1529835-2/ (accessed on 12 April 2018).

24. Official Journal of the Republic of Venezuela (Gaceta Oficial de la República de Venezuela). Decree No. 883 of the Republic of Venezuela, Pertaining to the Classification and Quality Control of Water Bodies and Contaminating Spillage. Federal Law Journal No. 5.021 of 18th December 1995. Available online: http://www.mvh.gob.ve/fabricadeinsumos27f/index.php?option=com_phocadownload&view=category&download=31:decreto-n-883-normas-para-la-clasificacion-y-el-control-de-la-calidad-de-los-cuerpos-de-agua-y-vertidos-o-efluentes-liquidos&id=11:leyes&Itemid=654 (accessed on 23 December 2018).

25. World Bank Group. *Environmental, Health, and Safety Guidelines for Onshore oil and Gas Development (English)*; IFC E&S; World Bank Group: Washington, DC, USA, 2007; Available online: http://documents.worldbank.org/curated/en/858751486372860509/Environmental-health-and-safety-guidelines-for-onshore-oil-and-gas-development (accessed on 26 March 2018).

26. World Bank Group. *Environmental, Health, and Safety Guidelines for Petroleum Refining (English)*; World Bank Group: Washington, DC, USA, 2016; Available online: http://documents.worldbank.org/curated/en/522801489581711256/Environmental-health-and-safety-guidelines-for-petroleum-refining (accessed on 26 March 2018).

27. Zheng, J.; Chen, B.; Thanyamanta, W.; Hawboldt, K.; Zhang, B.; Liu, B. Offshore produced water management: A review of current practice and challenges in harsh/Arctic environments. *Mar. Pollut. Bull.* **2016**, *104*, 7–19. [CrossRef] [PubMed]

28. Pan, L.; Chen, Y.; Chen, D.; Dong, Y.; Zhang, Z.; Long, Y. Oil removal in tight-emulsified petroleum waste water by flocculation. *IOP Conf. Ser. Mater. Sci. Eng.* **2018**, *392*, 042005. [CrossRef]

29. Zapata Acosta, K.; Carrasco-Marin, F.; Cortés, F.B.; Franco, C.A.; Lopera, S.H.; Rojano, B.A. Immobilization of P. stutzeri on Activated Carbons for Degradation of Hydrocarbons from Oil-in-Saltwater Emulsions. *Nanomaterials* **2019**, *9*, 500. [CrossRef] [PubMed]

30. Jamaly, S.; Giwa, A.; Hasan, S.W. Recent improvements in oily wastewater treatment: Progress, challenges, and future opportunities. *J. Environ. Sci.* **2015**, *37*, 15–30. [CrossRef] [PubMed]

31. Hu, G.; Li, J.; Zeng, G. Recent development in the treatment of oily sludge from petroleum industry: A review. *J. Hazard. Mater.* **2013**, *261*, 470–490. [CrossRef]

32. Khan, W.Z.; Najeeb, I.; Tuiyebayeva, M.; Makhtayeva, Z. Refinery wastewater degradation with titanium dioxide, zinc oxide, and hydrogen peroxide in a photocatalytic reactor. *Process Saf. Environ. Prot.* **2015**, *94*, 479–486. [CrossRef]

33. Chen, D.; Li, F.; Gao, Y.; Yang, M. Pilot Performance of Chemical Demulsifier on the Demulsification of Produced Water from Polymer/Surfactant Flooding in the Xinjiang Oilfield. *Water* **2018**, *10*, 1874. [CrossRef]

34. Karadag, D.; Köroğlu, O.E.; Ozkaya, B.; Cakmakci, M. A review on anaerobic biofilm reactors for the treatment of dairy industry wastewater. *Process Biochem.* **2015**, *50*, 262–271. [CrossRef]

35. Padaki, M.; Surya Murali, R.; Abdullah, M.S.; Misdan, N.; Moslehyani, A.; Kassim, M.A.; Hilal, N.; Ismail, A.F. Membrane technology enhancement in oil–water separation. A review. *Desalination* **2015**, *357*, 197–207. [CrossRef]

36. Subramani, A.; Jacangelo, J.G. Emerging desalination technologies for water treatment: A critical review. *Water Res.* **2015**, *75*, 164–187. [CrossRef] [PubMed]

37. Wang, P.; Chung, T.-S. Recent advances in membrane distillation processes: Membrane development, configuration design and application exploring. *J. Membr. Sci.* **2015**, *474*, 39–56. [CrossRef]

38. Alzahrani, S.; Mohammad, A.W. Challenges and trends in membrane technology implementation for produced water treatment: A review. *J. Water Process Eng.* **2014**, *4*, 107–133. [CrossRef]

39. Siyal, M.I.; Lee, C.-K.; Park, C.; Khan, A.A.; Kim, J.-O. A review of membrane development in membrane distillation for emulsified industrial or shale gas wastewater treatments with feed containing hybrid impurities. *J. Environ. Manag.* **2019**, *243*, 45–66. [CrossRef] [PubMed]

40. Mohammadtabar, F.; Khorshidi, B.; Hayatbakhsh, A.; Sadrzadeh, M. Integrated Coagulation-Membrane Processes with Zero Liquid Discharge (ZLD) Configuration for the Treatment of Oil Sands Produced Water. *Water* **2019**, *11*, 1348. [CrossRef]

41. Mohanakrishna, G.; Al-Raoush, R.I.; Abu-Reesh, I.M.; Aljaml, K. Removal of petroleum hydrocarbons and sulfates from produced water using different bioelectrochemical reactor configurations. *Sci. Total Environ.* **2019**, *665*, 820–827. [CrossRef]

42. Benito-Alcázar, C.; Vincent-Vela, M.; Gozálvez-Zafrilla, J.; Lora-García, J. Study of different pretreatments for reverse osmosis reclamation of a petrochemical secondary effluent. *J. Hazard. Mater.* **2010**, *178*, 883–889. [CrossRef]

43. Lu, M.; Wei, X. Treatment of oilfield wastewater containing polymer by the batch activated sludge reactor combined with a zerovalent iron/EDTA/air system. *Bioresour. Technol.* **2011**, *102*, 2555–2562. [CrossRef]

44. Hu, G.; Li, J.; Hou, H. A combination of solvent extraction and freeze thaw for oil recovery from petroleum refinery wastewater treatment pond sludge. *J. Hazard. Mater.* **2015**, *283*, 832–840. [CrossRef]

45. Zhao, S.; Huang, G.; Cheng, G.; Wang, Y.; Fu, H. Hardness, COD and turbidity removals from produced water by electrocoagulation pretreatment prior to Reverse Osmosis membranes. *Desalination* **2014**, *344*, 454–462. [CrossRef]

46. Dos Santos, E.V.; Rocha, J.H.B.; de Araújo, D.M.; de Moura, D.C.; Martínez-Huitle, C.A. Decontamination of produced water containing petroleum hydrocarbons by electrochemical methods: A minireview. *Environ. Sci. Pollut. Res.* **2014**, *21*, 8432–8441. [CrossRef] [PubMed]

47. Gargouri, B.; Gargouri, O.D.; Gargouri, B.; Trabelsi, S.K.; Abdelhedi, R.; Bouaziz, M. Application of electrochemical technology for removing petroleum hydrocarbons from produced water using lead dioxide and boron-doped diamond electrodes. *Chemosphere* **2014**, *117*, 309–315. [CrossRef] [PubMed]

48. Aljuboury, D.; Palaniandy, P.; Abdul Aziz, H.; Feroz, S. Treatment of petroleum wastewater by conventional and new technologies-A review. *Glob. Nest J.* **2017**, *19*, 439–452.

49. Jafarinejad, S. Treatment of Oily Wastewater. In *Petroleum Waste Treatment and Pollution Control*; Jafarinejad, S., Ed.; Butterworth-Heinemann: Oxford, UK, 2017; pp. 185–267, ISBN 978-0-12-809243-9.

50. Nasiri, M.; Jafari, I. Produced Water from Oil-Gas Plants: A Short Review on Challenges and Opportunities. *Periodica Polytech. Chem. Eng.* **2017**, *61*, 73–81. [CrossRef]

51. Nonato, T.C.M.; Alves, A.A.D.A.; Sens, M.L.; Dalsasso, R.L. Produced water from oil—A review of the main treatment technologies. *J. Environ. Toxicol.* **2018**, *2*, 23–27.

52. Ahmad, N.; Goh, P.; Abdul Karim, Z.; Ismail, A. Thin Film Composite Membrane for Oily Waste Water Treatment: Recent Advances and Challenges. *Membranes* **2018**, *8*, 86. [CrossRef] [PubMed]

53. Karhu, M.; Kuokkanen, T.; Rämö, J.; Mikola, M.; Tanskanen, J. Performance of a commercial industrial-scale UF-based process for treatment of oily wastewaters. *J. Environ. Manag.* **2013**, *128*, 413–420. [CrossRef]

54. Madaeni, S.S.; Eslamifard, M.R. Recycle unit wastewater treatment in petrochemical complex using reverse osmosis process. *J. Hazard. Mater.* **2010**, *174*, 404–409. [CrossRef]

55. Wan Ikhsan, S.N.; Yusof, N.; Aziz, F.; Nurasyikin, M. A review of oilfield wastewater treatment using membrane filtration over conventional technology. *MJAS* **2017**, *21*, 643–658. [CrossRef]

56. Masoudnia, K.; Raisi, A.; Aroujalian, A.; Fathizadeh, M. A hybrid microfiltration/ultrafiltration membrane process for treatment of oily wastewater. *Desalin. Water Treat.* **2014**, *55*, 1–12. [CrossRef]

57. Yan, L.; Ma, H.; Wang, B.; Mao, W.; Chen, Y. Advanced purification of petroleum refinery wastewater by catalytic vacuum distillation. *J. Hazard. Mater.* **2010**, *178*, 1120–1124. [CrossRef] [PubMed]

58. Taiwo, E.A.; Otolorin, J.A. Oil Recovery from Petroleum Sludge by Solvent Extraction. *Petrol. Sci. Technol.* **2009**, *27*, 836–844. [CrossRef]

59. SPSS. *Computer Software SPSS*; SPSS, Inc.: Chicago, IL, USA, 2015.

60. Zhang, S.; Wang, P.; Fu, X.; Chung, T.-S. Sustainable water recovery from oily wastewater via forward osmosis-membrane distillation (FO-MD). *Water Res.* **2014**, *52*, 112–121. [CrossRef] [PubMed]

61. El-Naas, M.H.; Alhaija, M.A.; Al-Zuhair, S. Evaluation of a three-step process for the treatment of petroleum refinery wastewater. *J. Environ. Chem. Eng.* **2014**, *2*, 56–62. [CrossRef]

62. Jeng, W.-L.; Huh, C.-A. A comparison of sedimentary aliphatic hydrocarbon distribution between the southern Okinawa Trough and a nearby river with high sediment discharge. *Estuar. Coast. Shelf Sci.* **2006**, *66*, 217–224. [CrossRef]

63. Mijaylova Nacheva, P.; Birkle, P.; Ramírez Camperos, E.; Sandoval Yoval, L. Tratamiento de aguas de la desalación del petróleo para su aprovechamiento en inyección al subsuelo. *Revista AIDIS de Ing y Cienc Amb Invest des y prac* **2007**, *1*, 1–16. [CrossRef]

64. Cortes, J.E.; Rincon, J.M.; Jaramillo, J.M.; Philp, R.P.; Allen, J. Biomarkers and compound-specific stable carbon isotope of n-alkanes in crude oils from Eastern Llanos Basin, Colombia. *J. S. Am. Earth Sci.* **2010**, *29*, 198–213. [CrossRef]

65. Peters, K.; Walters, C.; Moldowan, J. *The Biomarker Guide: Biomarkers and Isotopes in the Environment and Human History*, 2nd ed.; Cambridge University Press: New York, NY, USA; Cambridge, UK, 2005; Volume 1, p. 471, ISBN 978-05-1152-486-8.

66. Chavan, A.; Mukherji, S. Treatment of hydrocarbon-rich wastewater using oil degrading bacteria and phototrophic microorganisms in rotating biological contactor: Effect of N:P ratio. *J. Hazard. Mater.* **2008**, *154*, 63–72. [CrossRef]

67. Koo, C.H.; Mohammad, A.W. Recycling of oleochemical wastewater for boiler feed water using reverse osmosis membranes—A case study. *Desalination* **2011**, *271*, 178–186. [CrossRef]

68. Yuliwati, E.; Ismail, A.; Lau, W.; Ng, B.; Mataram, A.; Kassim, M. Effects of process conditions in submerged ultrafiltration for refinery wastewater treatment: Optimization of operating process by response surface methodology. *Desalination* **2012**, *287*, 350–361. [CrossRef]

69. Al-Deffeeri, N.S. Chemical analysis of distilled water: A case study. *Desalin. Water Treat.* **2013**, *51*, 1936–1940. [CrossRef]

70. Igunnu, E.T.; Chen, G.Z. Produced water treatment technologies. *Int. J. Low-Carbon Technol.* **2012**, *9*, 157–177. [CrossRef]

MDPI

St. Alban-Anlage 66

4052 Basel

Switzerland

Tel. +41 61 683 77 34

Fax +41 61 302 89 18

www.mdpi.com

Water Editorial Office

E-mail: water@mdpi.com

www.mdpi.com/journal/water

www.ingramcontent.com/pod-product-compliance
Lightning Source LLC
Chambersburg PA
CBHW051906210326
41597CB00033B/6045